DEPARTMENT OF THE NAVY
OFFICE OF THE CHIEF OF NAVAL OPERATIONS
2000 NAVY PENTAGON
WASHINGTON, DC 20350-2000

MEMORANDUM FOR DISTRIBUTION

Enclosure (1) Report on the Collision between USS FITZGERALD (DDG 62) and Motor Vessel ACX CRYSTAL
Enclosure (2) Report on the Collision between USS JOHN S MCCAIN (DDG 56) and Motor Vessel ALNIC MC

The collisions were avoidable between USS FITZGERALD (DDG 62) and Motor Vessel ACX CRYSTAL, and between USS JOHN S MCCAIN (DDG 56) and Motor Vessel ALNIC MC. Three U.S. Navy investigations concerning each of these incidents are complete. Command and Admiralty investigations in each case retain legal privilege to protect the interests of the United States Government in future litigation. The third investigation for each incident, termed the Line of Duty Investigation (LODI), is not under legal privilege as its purpose is to determine that Sailors perished in the line of duty and thus enable their beneficiaries to receive appropriate compensation. Collisions at sea between U.S. registered vessels and foreign registered vessels are also subject to an additional investigation, a Marine Casualty Investigation, conducted independently on behalf of the National Transportation Safety Board (NTSB) by the United States Coast Guard (USCG). These investigations are ongoing in each case the results of each will be published by the NTSB when complete.

As Chief of Naval Operations, I have determined to retain the legal privilege that exists with the command Admiralty investigations in order to protect the legal interests of the United States Government and the families of those Sailors who perished. At the same time, it is paramount that the Navy be transparent as to the causes and lessons learned to the families of those Sailors, the Congress and the American people, and to make every effort to ensure these types of tragedies do not happen again. With these competing interests at hand, I authorized the preparation and release of reports on each collision, enclosed with this memorandum.

These collisions, along with other similar incidents over the past year, indicated a need for the Navy to undertake a review of wider scope to better determine systemic causes. The Navy's Comprehensive Review of Surface Fleet Incidents, completed on 23 October 2017, represents the results of this effort. This review is enclosed with this memorandum, and represents a summary of significant actions needed to fix the larger problems and their causes leading up to these incidents.

Finally, the Navy has an obligation to protect the privacy of individuals involved in these incidents to the extent that it is possible. With legal, privacy and transparency concerns in mind, the enclosures to this memorandum provide the Navy's findings, conclusions and actions moving forward.

Table of Contents

ENCLOSURE (1)

REPORT ON THE COLLISION BETWEEN USS FITZGERALD (DDG 62)
AND MOTOR VESSEL ACX CRYSTAL

23 OCT 2017

1. EXECUTIVE SUMMARY - USS FITZGERALD

1.1 Introduction

USS FITZGERALD collided with Motor Vessel ACX CRYSTAL on 17 June 2017 in the waters of Sagami Wan in vicinity of the approaches to Tokyo Wan.

FITZGERALD is an Arleigh Burke Class Destroyer commissioned in 1995 and homeported in Yokosuka, Japan, as part of the Forward Deployed Naval Forces and Carrier Strike Group FIVE. Approximately 300 Sailors serve aboard FITZGERALD. FITZGERALD is 505 feet in length and carries a gross tonnage of approximately 9000 tons.

Figure 1 illustrates the relative sizes of the vessels. ACX CRYSTAL (CRYSTAL) is a Philippines flagged container ship built in 2008. CRYSTAL is 728 feet long with a gross tonnage of approximately 29,000 tons.

The collision between FITZGERALD and CRYSTAL resulted in the deaths of seven U.S. Sailors due to impact with FITZGERALD's berthing compartments, located below the waterline of the ship. CRYSTAL suffered no fatalities. U.S. Sailor fatalities were:

GMSN Kyle Rigsby of Palmyra, Virginia, 19 years old.

YN2 Shingo Alexander Douglass, of San Diego, California, 25 years old.

FC1 Carlos Victor Ganzon Sibayan of Chula Vista, California, 23 years old.

PSC Xavier Alec Martin of Halethorpe, Maryland, 24 years old.

STG2 Ngoc Turong Huynh of Oakville, Connecticut, 25 years old.

GM1 Noe Hernandez of Weslaco, Texas, 26 years old.

FCC Gary Rehm, Jr., of Elyria, Ohio, 37 years old.

1.2 Summary of Findings

The Navy determined that numerous failures occurred on the part of leadership and watchstanders as follows:

Failure to plan for safety.
Failure to adhere to sound navigation practice.
Failure to execute basic watch standing practices.
Failure to properly use available navigation tools.
Failure to respond deliberately and effectively when in extremis.

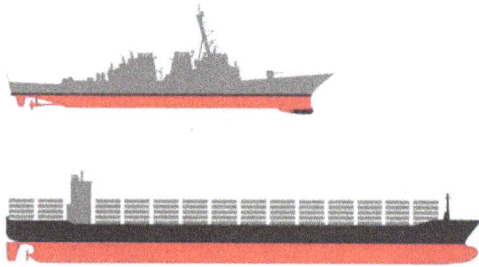

Figure 1 – Relative size of the USS Fitzgerald

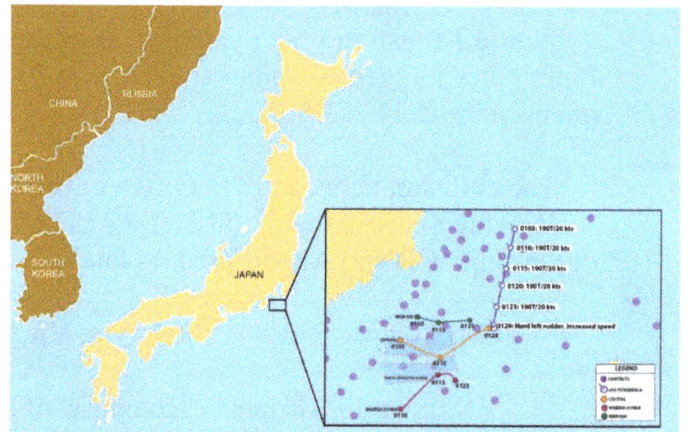

Figure 2 – Illustration Map of Approximate Collision Location

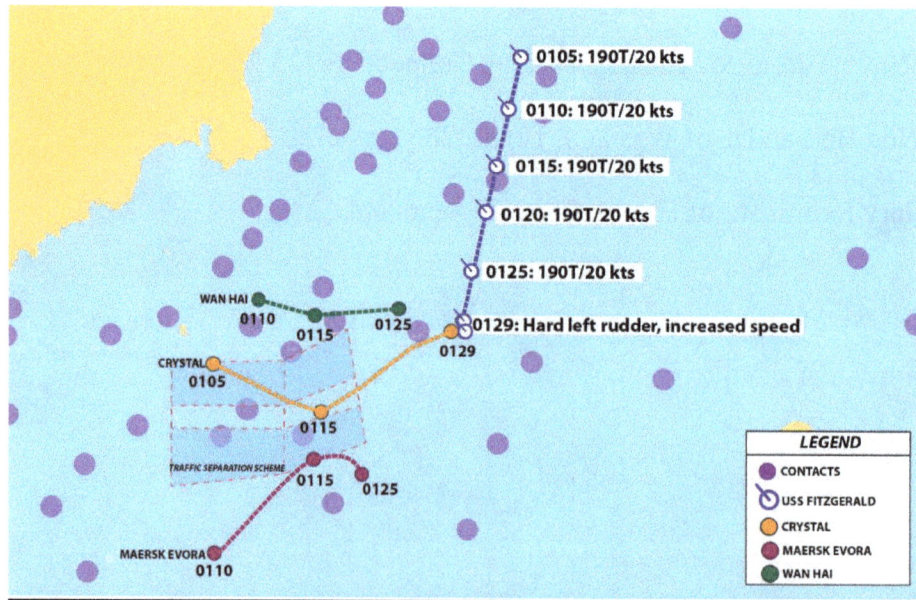

Figure 3 – Illustration Map of Approximate Collision Location

2. DESCRIPTION OF EVENTS

2.1 Background

On the morning of 16 June 2017, FITZGERALD departed the homeport of Yokosuka, Japan for routine operations. The weather was pleasant with unlimited visibility and calm seas. After a long day of training evolutions and equipment loading operations, FITZGERALD proceeded southwest on a transit to sea from the Sagami Wan operating area at approximately 2300.

FITZGERALD was operating by procedures established for U.S. Navy surface ships when operating at sea before sunrise, including being at "darkened ship." "Darkened ship" means that all exterior lighting was off except for the navigation lights that provide identification to other vessels, and all interior lighting was switched to red instead of white to facilitate crew rest. The ship was in a physical posture known as "Modified ZEBRA," meaning that all doors inside the ship, and all hatches, which are openings located on the floor between decks, at the main deck and below were shut to help secure the boundaries between different areas of the ship in case of flooding or fire. Watertight scuttles on the hatches (smaller circular openings that can be opened or closed independently of the hatch) were left open in order to allow easy transit between spaces.

By 0130 hours on 17 June 2017, the approximate time of the collision, FITZGERALD was approximately 56 nautical miles to the southwest of Yokosuka, Japan, near the Izu Peninsula within sight of land and continuing its transit outbound. The seas were relatively calm at 2 to 4 feet. The sky was dark, the moon was relatively bright, and there was scattered cloud cover and unrestricted visibility.

2.2 Events Leading to the Collision

At approximately 2300 local Japan time, both the Commanding Officer and Executive Officer departed the bridge, the area from which watchstanders drive the ship. As the FITZGERALD proceeded past Oshima Island the shipping traffic increased and remained moderately dense thereafter until the collision. By 0100, FITZGERALD approached three merchant vessels from its starboard, or right side, forward. These vessels were eastbound through the Mikomoto Shima Vessel Traffic Separation Scheme. Traffic separation schemes are established by local authorities in approaches to ports throughout the world to provide ships assistance in separating their movements when transiting to and from ports. The closest point of approach of these vessels and the FITZGERALD was minimal with each presenting a risk of collision.

In accordance with the International Rules of the Nautical Road, the FITZGERALD was in what is known as a crossing situation with each of the vessels. In this situation, FITZGERALD was obligated to take maneuvering action to remain clear of the other three, and if possible, avoid crossing ahead. In the event FITZGERALD did not exercise this obligation, the other vessels were obligated to take early and appropriate action through their own independent maneuvering action. In the 30 minutes leading up to the collision, neither FITZGERALD nor CRYSTAL took such action to reduce the risk of collision until approximately one minute prior to the collision. FITZGERALD maintained a constant course of 190 degrees at 20 knots of speed.

In the several minutes before collision, the Officer of the Deck, the person responsible for safe navigation of the ship, and the Junior Officer of the Deck, an officer placed to assist, discussed the relative positioning of the vessels, including CRYSTAL and whether or not action needed taken to avoid them. Initially, the Officer of the Deck intended to take no action, mistaking CRYSTAL to be another of the two vessels with a greater closest point of approach. Eventually, the Officer of the Deck realized that FITZGERALD was on a collision course with CRYSTAL, but this recognition was too late. CRYSTAL also took no action to avoid the collision until it was too late.

The Officer of the Deck, the person responsible for safe navigation of the ship, exhibited poor seamanship by failing to maneuver as required, failing to sound the danger signal and failing to attempt to contact CRYSTAL on Bridge to Bridge radio. In addition, the Officer of the Deck did not call the Commanding Officer as appropriate and prescribed by Navy procedures to allow him to exercise more senior oversight and judgment of the situation.

The remainder of the watch team on the bridge failed to provide situational awareness and input to the Officer of the Deck regarding the situation. Additional teams in the Combat Information Center (CIC), an area on where tactical information is fused to provide maximum situational awareness, also failed to provide the Officer of the Deck input and information.

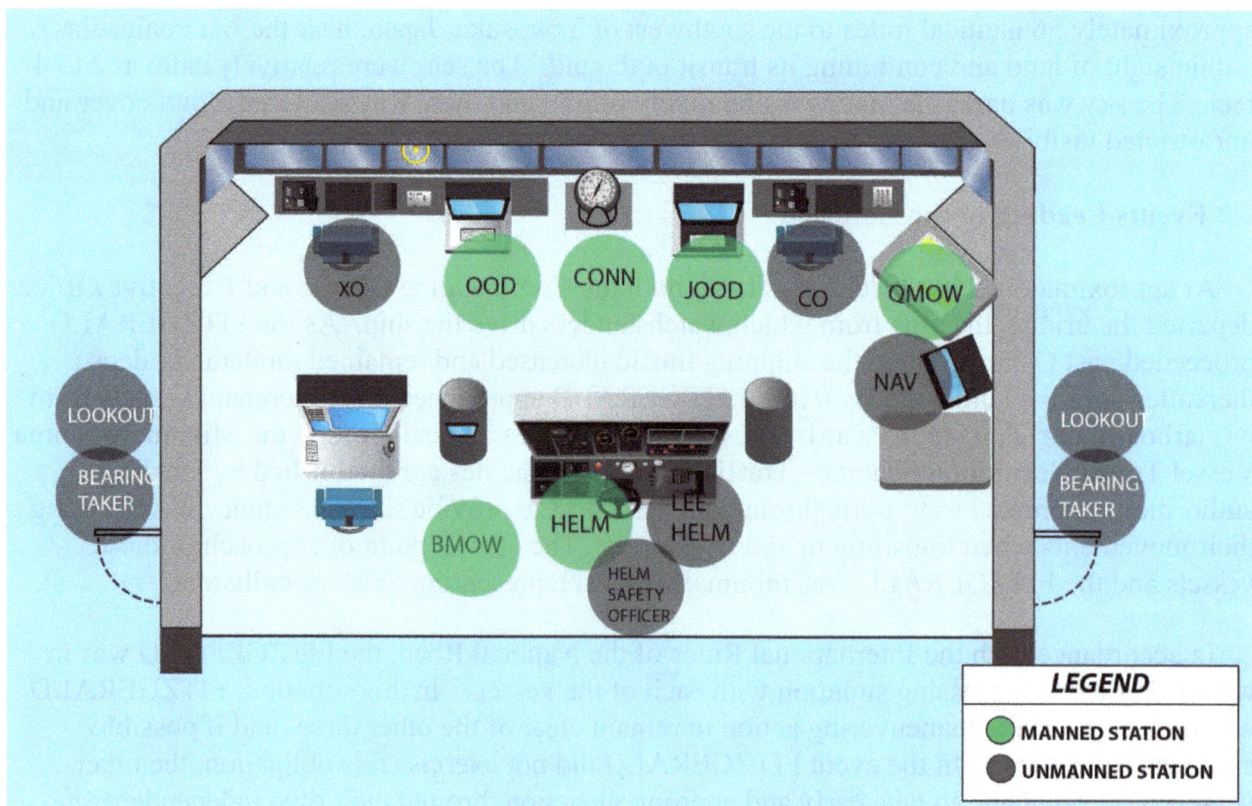

Figure 4 – Bridge Schematic of FITZGERALD

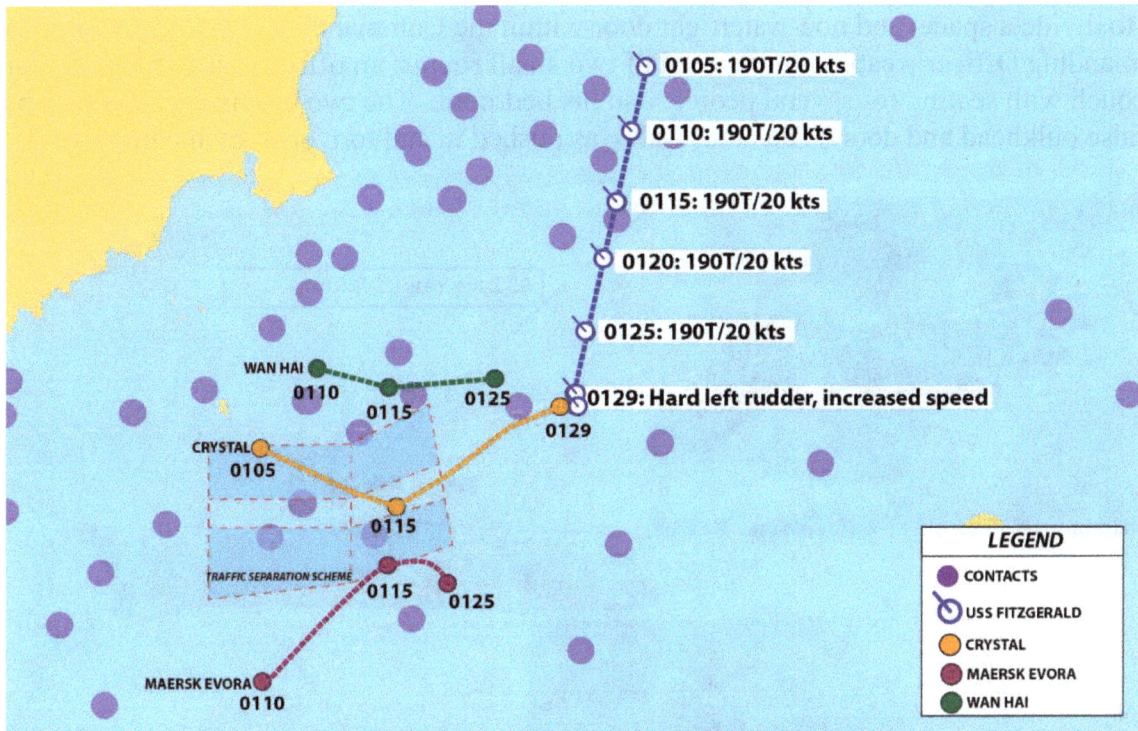

Figure 5 – Illustration of Approximate Collision Location

3. IMPACT OF THE COLLISION

The port (left) side of CRYSTAL's bow, near the top where the anchor hangs, struck FITZGERALD's starboard (right) side above the waterline. CRYSTAL's bulbous bow, under the water, struck FITZGERALD on the starboard side just forward of the middle part of the ship. CRYSTAL's bulbous bow struck the starboard access trunk, an entry space that opens into Berthing 2 through a non-watertight door.

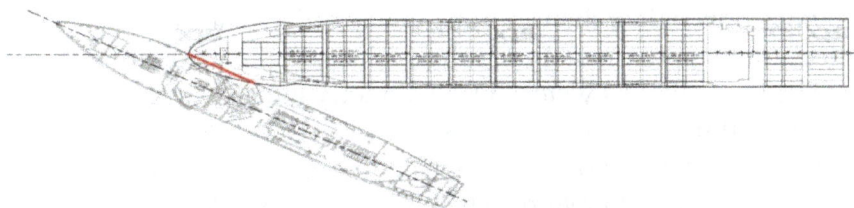

Figure 6 – Diagram of Approximate Collision Geometry

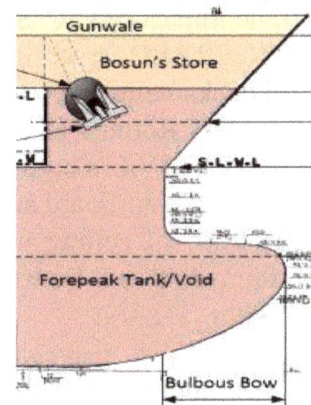

Figure 7 - Depiction of a Bow and Bulbous Bow

8

The impact of the top of CRYSTAL's bow above the waterline crushed the false bulkhead (a non-structural steel wall that is the shipboard equivalent of a non-load bearing wall in a house, used to divide a space) and non-watertight door within the Commanding Officer's cabin. The Commanding Officer's cabin is composed of two small rooms: an office area with a desk, table, and couch with seating for several people; and his bedroom. The two rooms are separated by this false bulkhead and door. The outer wall was pushed in and torn open by the impact.

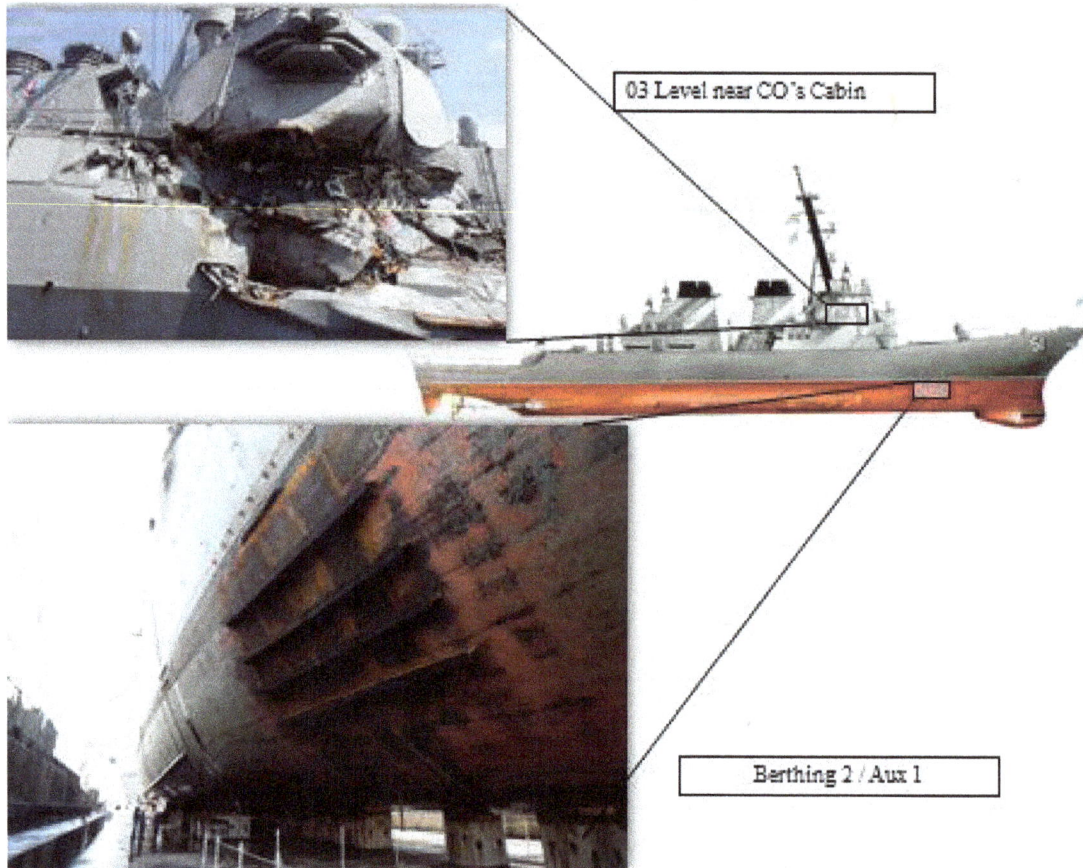

Figure 8 - Starboard side of FITZGERALD
Inset Above: Damage to FITZGERALD above the waterline at the 03 Level near CO's Cabin
Inset below: FITZGERALD in dry dock with patch over the 13ft x 17ft hole at Berthing 2/AUX 1

The impact of CRYSTAL's bulbous bow below the waterline punctured the side of FITZGERALD, creating a hole measuring approximately 13ft by 17ft, spanning the second and third decks below the main deck. The hole allowed water to flow directly into Auxiliary Machinery Room 1 (AUX 1) and the Berthing 2 starboard access trunk. The force of impact from CRYSTAL's bulbous bow and resulting flood of water pushed the non-water tight door between the starboard access trunk and Berthing 2 inward. The wall supporting this door pulled away from the ceiling and bent to a near-90 degree angle. As a result, nothing separated Berthing 2 from the onrushing sea, allowing a great volume of water to enter Berthing 2 very quickly.

The impact at the moment of collision caused FITZGERALD to list (tilt) a reported 14

9

degrees to port. FITZGERALD then settled into a 7 degree starboard list as the sea flooded into Berthing 2 through the starboard access trunk and weighted the ship deeper into the water on the starboard side.

Water poured into the ship from the hole in the hull and flooded into spaces directly connected to areas near the hole or not separated by a watertight barrier, including Berthing 2 and associated spaces; AUX 1 and associated spaces; and other spaces forward of the hole. Other spaces were flooded due to cross flooding, which is the flooding of spaces that are connected to damaged spaces and that have the ability to be isolated with a water tight barrier, but could not be sealed off in time due to the rapid flooding caused by the large hole in the side of the ship. Additionally, spaces were partially flooded due to a ruptured fire main (large seawater pipes that provide water for fighting fires) and ruptured Aqueous Film-Forming Foam (a form of firefighting agent) piping.

Multiple spaces suffered structural damage: Commanding Officer's Cabin, Stateroom, and Bathroom; Officer Stateroom; Berthing 2 Starboard Access Trunk; Auxiliary Compartment 1; Repair Locker Number 2 passageway; Combat Information Center passageway; multiple Radar and Radar Array rooms; multiple fan rooms; Combat Systems Maintenance Central airlock and ladder-well; Electronic Workshop Number 1; and the Starboard Break.

The collision resulted in a loss of external communications and a loss of power in the forward portion of the ship. Following the collision, FITZGERALD changed the lighting configuration at the mast to one red light over another red light, known as "red over red," the international lighting scheme that indicates a ship that is "not under command." Under International Rules of the Nautical Road, this signifies that, due to an exceptional circumstance such as loss of propulsion or steering, a vessel is unable to maneuver as required.

All U.S. Navy ships are designed to withstand and recover from damage due to fire, flooding, and other damage sustained during combat or other emergencies. Each ship has a Damage Control Assistant, working under the Engineering Officer, in order to establish and maintain an effective damage control organization. The Damage Control Assistant oversees the prevention and control of damage including control of stability, list, and trim due to flooding (maintaining the proper level of the ship from side to side and front and back), coordinates firefighting efforts, and is also responsible for the operation, care and maintenance of the ship's repair facilities. The DCA ensures the ship's repair party personnel are properly trained in damage control procedures including firefighting, flooding and emergency repairs. The Damage Control Assistant is assisted by the Damage Control Chief, a chief petty officer specializing in Damage Control. The officer in charge of damage control efforts, the Damage Control Assistant, called away General Quarters to notify the crew to commence damage control efforts.

General Quarters is a process whereby the crew reports to pre-assigned stations and duties in the event of large casualties such as flooding. General Quarters is announced by an alarm that sounds throughout the ship to alert the crew of an emergency situation or potential combat operations. All crewmembers are trained to report to their General Quarters watch station and to set a higher condition of material readiness against fire, flooding, or other damage. This involves securing additional doors, hatches, scuttles, valves and equipment to isolate damage and prepare for combat. Navy crews train on Damage Control continuously with drills being run in port and underway frequently to prepare the teams for damage to equipment and spaces. During any emergency condition (fire, flooding, combat operations), the Damage Control Assistant coordinates and supervises all damage control efforts from one of the three Damage Control Repair Lockers.

Damage Control Repair Lockers are specialized spaces stationed throughout the ship filled with repair equipment and manned during emergencies with teams of about 20 personnel trained to respond to casualties. There are three repair lockers on the FITZGERALD: Repair Locker 2, Repair Locker 5, and Repair Locker 3. Repair Locker 2 covers the forward part of the ship, Repair Locker 5 covers the engineering spaces and Repair Locker 3 covers the aft part of the ship. Each locker is maintained with similar equipment. Personnel assigned to repair lockers are trained and qualified to respond to, and repair damage from, a variety of sources with a specific focus on fire and flooding. Each repair locker can act independently but is also designed to support the others and can take over the responsibilities for any locker if damage prevents that locker's use. The repair lockers are normally unmanned unless the ship sets a condition of higher readiness like General Quarters when they would be manned within minutes.

The Damage Control Assistant and Damage Control Chief took control of damage control actions immediately after the collision, organizing General Quarters and ship wide efforts. Reports were received of a 12 foot by 12 foot hole in Auxiliary Compartment 1 and flooding. Reports were also received of a 3 foot by 5 foot hole and flooding in the right side (starboard) passage to Berthings 1 and 2. These two reports were determined to be differing accounts of the same hole caused by the collision that had spread to two spaces. The hole was later determined to be 17ft by 13ft.

Based on reports of flooding in Berthing 2 and Auxiliary Compartment 1, supervisors attempted to direct the Repair Locker 2 team, the closest to the report location, in place to combat the flooding. However, due to location of the flooding, the supervisors could not communicate with Repair Locker 2. Efforts were reassigned to Repair Locker 5 as communications were established on internal communications networks still operable, and hand-held radios and sound powered phones, which are phones that do not require electrical power to operate.

Damage control parties used eductors to remove water from flooded spaces. Eductors use a jet of water, typically supplied from the ship's fire-main piping system, to remove water from spaces. FITZGERALD also used three onboard pumps to remove water from the ship. Two of the pumps functioned as designed and a third seized and was inoperable for the duration of the recovery efforts.

Berthing 1 was partially flooded to five feet by water entering from the hole created by the impact in Berthing 2 and AUX 1. Damage control parties tried to limit and reduce the flooding with eductors, but with no watertight hatch or door between Berthing 2 and Berthing 1 on the starboard side to act as a barrier against the progressive flooding, water passed freely into Berthing 1 and undermined dewatering efforts. Berthing 1 had water in the space until after the ship returned to Yokosuka and entered dry dock on 11 July. Dewatering of Berthing 2 was completed only after the ship entered dry dock. Auxiliary Compartment 1 was dewatered using an additional pump once a patch was installed in Yokosuka.

The Starboard Passageway was partially flooded due to the ruptured fire-main and Aqueous Film Forming Foam (a form of firefighting agent) piping as a direct result of the impact. 1.5 feet of water was reported on deck with 3 feet of firefighting foam. Damage Control Parties used two eductors to effectively counter the flooding. They also used shoring, a material designed to reinforce a structural defect, along the passageway to mitigate the effects of the collision.

Radio Central was partially flooded as a result of flooding in space ins close proximity. The crew effectively combatted flooding with stuffing compound used in cable way repairs and then dewatered the space using buckets and mops. Main Engine Room 1 had minor flooding due to proximity flooding from Auxiliary Compartment 1's Waste Water Tank and Oily Waste Tank. An eductor was aligned to effectively dewater the space and the water level remained below three feet.

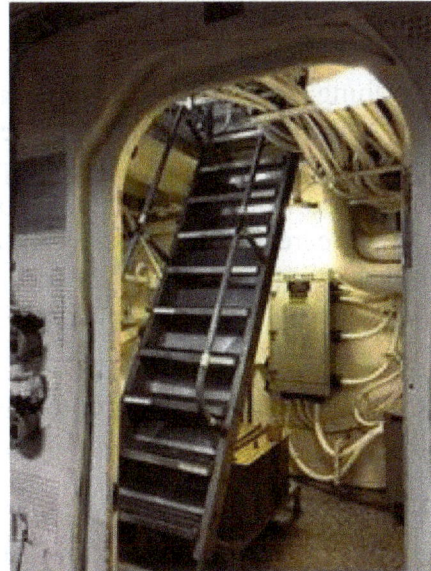

Figure 9 – Non-watertight door frame from Berthing 2 to the ladder going up to Berthing 1 Access Trunk. There is no hatch at the top of this ladder to prevent water from flooding up into Berthing 1.

Sonar Control, Sonar Control Fan Room, Sonar Control Passageway, Sonar Administration Office, and Combat Systems Equipment Room 1 were partially flooded with five feet of water on the deck. Subsequent damage control efforts effectively dewatered the space using an eductor and pump.

The Forward 400Hz, Fan Room, and Power Conversion Room experienced both flooding and white smoke. Progressive flooding through the vents in AUX 1 resulted in two feet of water on the deck, which was engaged using eductors, buckets, and swabs. Dewatering efforts continued while water continued to flow in, maintaining one foot depth of water or less.

As Sailors reported to their damage control duties and undertook efforts, departments began to account for missing Sailors. Reports were received that three Sailors were trapped in Sonar Control as a result of the collision. Realizing there was flooding in the spaces above them, the Sailors in Sonar Control radioed for assistance. A team was sent in but initial attempts to reach them were unsuccessful because the passageway was completely obstructed due to damage. This was also one of the areas that suffered cross-flooding through deck drains that could not be secured before flooding advanced. The team reached the escape hatch above the Sonar Control space, which was topped with a few inches water. They went through the hatch and were able to assist the Sailors trapped inside at approximately 0215.

At 0225, two Sailors were identified as unaccounted for in the Combat Systems Department. At approximately 0316, four Sailors were reported as unaccounted for. At 0540, a final, accurate all-hands accounting was received in which seven Sailors remained missing.

3.1 Impact to Berthing 2

Berthing 2 is a crew area containing 42 racks (beds) and spans from one side of the ship across to the other side two decks below the main deck of the ship. It includes its own bathroom, a shower and bathroom space accessed via a non-watertight door. It also includes a lounge filled with sofas, chairs, a table, and a television where the crew in the berthing can relax and recreate. The space is approximately 29 feet long, approximately 40 feet across, and with ceilings approximately 10 feet high. The racks are stacked with a top, middle, and bottom rack, each with a mattress and privacy curtains. Figure 7 provides an example of how the racks would have been configured in Berthing 2.

Figure 10 – Berthing 2 Layout Diagram (facing aft)

Berthing 2 has three ways out (egress points), of which two are on the port side. The first

port side egress is up a ladder through a watertight hatch with a watertight scuttle at the top of the ladder. Figures 8 and 9 show this egress point on a ship of the same class as

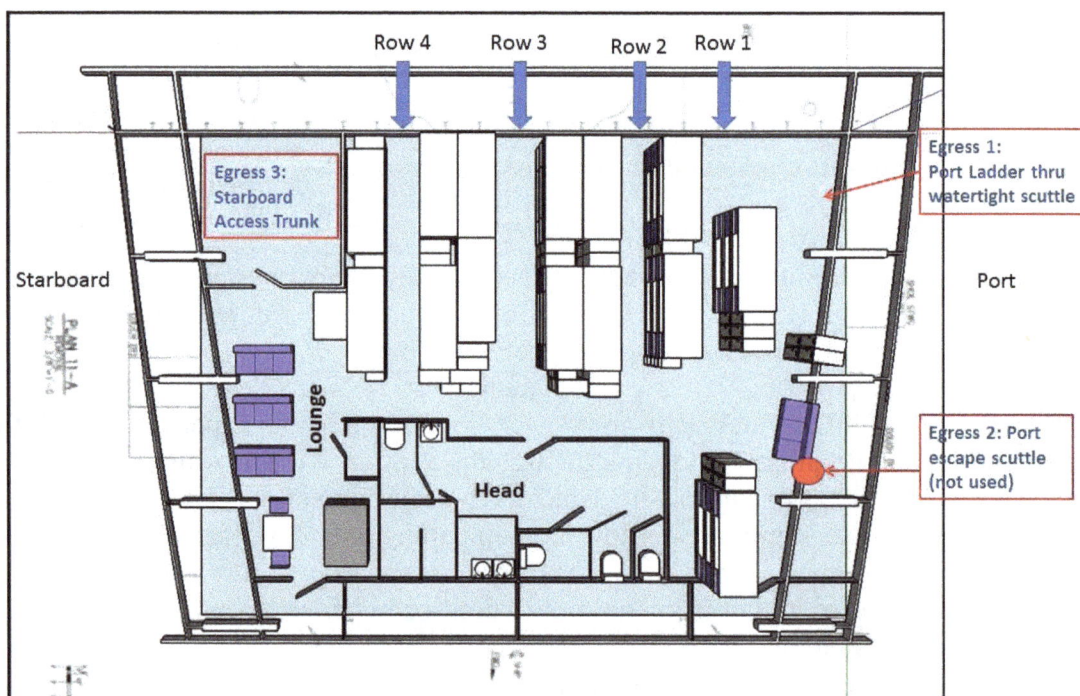

FITZGERALD. The second port side egress is an escape scuttle in the ceiling that leads directly into Berthing 1. Figure 10 shows this egress point on a ship of the same class as

FITZGERALD. This escape scuttle is usually in the down position, as it would become a trip hazard in Berthing 1 if left in the up position.

Figure 11 – Sample Berthing 2 Starboard Side Egress – Ladder Up

Figure 12 – Sample Berthing 2 Starboard Side Egress – Ladder Up

Figure 13 – Sample Berthing 2 Starboard Side Egress – Scuttle Down to Forward IC (no outlet)

The third egress point is on the starboard side, where CRYSTAL's bulbous bow struck

FITZGERALD. A non-watertight door leads from Berthing 2 to the starboard access trunk. In that trunk, a ladder leads up one deck into a space just outside of Berthing 1. Figures 11 through 13 are examples of the starboard side egress on a ship of the same class as FITZGERALD. There is no hatch separating the starboard access trunk outside Berthing 2 and the space above it. Also in this starboard access trunk is a hatch and watertight scuttle connected to a ladder going down to the Forward Interior Communications (FWD IC) space. FWD IC is not manned underway and has no other exits.

Of the 42 Sailors assigned to Berthing 2, at the time of collision, five were on watch and two were not aboard. Of the 35 remaining Sailors in Berthing 2, 28 escaped the flooding. Seven Sailors perished.

Some of the Sailors who survived the flooding in Berthing 2 described a loud noise at the time of impact. Other Berthing 2 Sailors felt an unusual movement of the ship or were thrown from their racks. Other Berthing 2 Sailors did not realize what had happened and remained in their racks. Some of them remained asleep. Some Sailors reported hearing alarms after the collision, while others remember hearing nothing at all.

Seconds after impact, Sailors in Berthing 2 started yelling, "Water on deck!" and, "Get out!" One Sailor saw another knocked out of his rack by water. Others began waking up shipmates who had slept through the initial impact. At least one Sailor had to be pulled from his rack and into the water before he woke up. Senior Sailors checked for others that might still be in their racks.

The occupants of Berthing 2 described a rapidly flooding space, estimating later that the space was nearly flooded within a span of 30 to 60 seconds. By the time the third Sailor to leave arrived at the ladder, the water was already waist deep. Debris, including mattresses, furniture, an exercise bicycle, and wall lockers, floated into the aisles between racks in Berthing 2, impeding Sailors' ability to get down from their racks and their ability to exit the space. The ship's 5 to 7 degree list to starboard increased the difficulty for Sailors crossing the space from

14

the starboard side to the port side. Many of the Sailors recall that the battle lanterns were illuminated. Battle lanterns turn on when power to an electrical circuit is out or when turned on manually. The yellow boxes hanging from the ceiling in Figure 14 are battle lanterns.

Sailors recall that after the initial shock, occupants lined up in a relatively calm and orderly manner to climb the port side ladder and exit through the port side watertight scuttle. Figure 14 provides an example of the route Sailors would have taken from their racks to the port side watertight scuttle on a ship of the same class as FITZGERALD. They moved along the blue floor and turned left at the end to access the ladder. Figure 14 provides an example and sense of scale. Even though the Sailors were up to their necks in water by that point, they moved forward slowly and assisted each other. One Sailor reported that FC1 Rehm pushed him out from under a falling locker. Two of the Sailors who already escaped from the main part of Berthing 2 stayed at the bottom of the ladder well (see Figure 8) in order to help their shipmates out of the berthing area.

The door to the Berthing 2 head (bathrooms and showers) was open and the flooding water dragged at least one person into this area. Exiting from the head during this flood of water was difficult and required climbing over debris.

As the last group of Sailors to escape through the port side watertight scuttle arrived at the bottom of the ladder, the water was up to their necks. The two Sailors who had been helping people from the bottom of the ladder were eventually forced to climb the ladder as water reached the very top of the Berthing 2 compartment. They continued to assist their shipmates as they climbed, but were eventually forced by the rising water to leave Berthing 2 through the watertight scuttle themselves. Before climbing the ladder, they looked through the water and did not see any other Sailors. Once through the watertight scuttle and completely out of the Berthing 2 space (on the landing outside Berthing 1) they continued to search, reaching

Figure 14 – Sample Berthing 2 View from Row 3 of racks to the port side (open door on right leading to head)

into the dark water to try to find anyone they could. From the top of the ladder, these two Sailors were able to pull two other Sailors from the flooded compartment. Both of the rescued Sailors were completely underwater when they were pulled to safety.

The last Sailor to be pulled from Berthing 2 was in the bathroom at the time of the collision and a flood of water knocked him to the deck (floor). Lockers were floating past him and he scrambled across them towards the main berthing area. At one point he was pinned between the lockers and the ceiling of Berthing 2, but was able to reach for a pipe in the ceiling to pull himself free. He made his way to the only light he could see, which was coming from the port side watertight scuttle. He was swimming towards the watertight scuttle when he was pulled from the water, red-faced and with bloodshot eyes. He reported that when taking his final breath before being saved, he was already submerged and breathed in water.

After the last Sailor was pulled from Berthing 2, the two Sailors helping at the top of the port side watertight scuttle noticed water coming into the landing from Berthing 1. They remained in case any other Sailors came to the ladder. Again, one of the Sailors stuck his arms through the watertight scuttle and into the flooded space to try and find any other Sailors, even as the area around him on the landing outside of Berthing 1 flooded. Berthing 1, with no watertight door between it and the landing, began to flood.

Another Sailor returned with a dogging wrench, a tool used to tighten the bolts, on the hatch to stave off flooding from the sides of the hatch. The three Sailors at the top of the ladder yelled into the water-filled space below in an attempt to determine if there was anyone still within Berthing 2. No shadows were seen moving and no response was given.

Water began shooting up and out of the watertight scuttle into the landing. Finding no other Sailors, they tried to close the watertight scuttle to stop the flood of water. The force of the water through the hatch prevented closing the watertight scuttle between Berthing 2 and Berthing 1. The scuttle was left partially open. They then climbed the ladder to the Main deck (one level up from the Berthing 1 landing), and secured the hatch and scuttle between Berthing 1 and the Main deck. In total, 27 Sailors escaped Berthing 2 from the port side ladder.

One Sailor escaped via the starboard side of Berthing 2. After the collision, this Sailor tried to leave his rack, the top rack in the row nearest to the starboard access trunk, but inadvertently kicked someone, so he crawled back into his rack and waited until he thought everyone else would be out of the Berthing 2. When he jumped out of his rack a few seconds later, the water was chest high and rising, reaching near to the top of his bunk.

After leaving his rack, the Sailor struggled to reach the starboard egress point through the lounge area. He moved through the lounge furniture and against the incoming sea. Someone said, "go, go, go, it's blocked," but he was already underwater. He was losing his breath under the water but found a small pocket of air. After a few breaths in the small air pocket, he eventually took one final breath and swam. He lost consciousness at this time and does not remember how he escaped from Berthing 2, but he ultimately emerged from the flooding into Berthing 1, where he could stand to his feet and breathe. He climbed Berthing 1's egress ladder, through Berthing 1's open watertight scuttle and collapsed on the Main Deck. He was the only Sailor to escape through the starboard egress point.

The flooding of Berthing 2 resulted in the deaths of seven FITZGERALD Sailors. The racks of these seven Sailors were located in Rows 3 and 4, the area closest to the starboard access trunk and egress point and directly in the path of the onrushing water, as depicted in Figure 15.

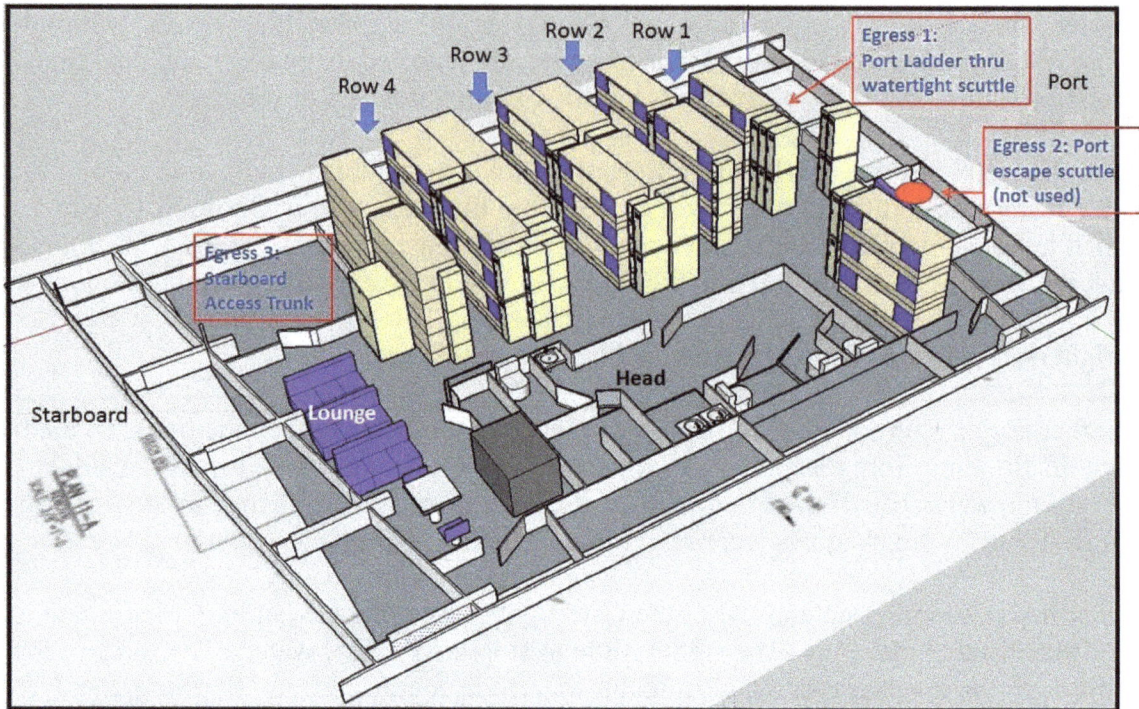

Figure 15 – Berthing 2 Layout of Racks and Lockers (facing aft)

After escaping Berthing 2, Sailors went to various locations. Some assembled on the mess decks to treat any injuries and pass out food and water. Others went to their General Quarters (GQ) stations to assist with damage control efforts. Another Sailor went to the bridge to help with medical assistance. One Sailor later took the helm and stood a 15-hour watch in aft steering after power was lost forward.

3.2 Other Rescue and Medical Efforts

The Sailor who escaped from the starboard egress point was in shock and was quickly moved to the administrative office for medical treatment. Due to the severity of his injuries, he was medically evacuated to U.S. Naval Hospital Yokosuka (USNHY) via helicopter at approximately 0915 on 17 June 2017. He was treated for near drowning, seawater aspiration, traumatic brain injury, and scalp and ankle lacerations and was released on 19 June 2017.

One Officer was injured in the collision. He was not in his stateroom at the time of the collision, but returned to it and assisted his roommate, who was trapped inside his rack by the force of the collision. The Officer's stateroom immediately adjoins the Commanding Officer's stateroom and just aft of the point of collision. He suffered a traumatic brain injury, lost consciousness, and had cuts and bruises to his head. Due to the severity of his injuries, he was medically evacuated to USNHY via helicopter at approximately 0915 on 17 June 2017. He was treated and released on 18 June 2017.

The Commanding Officer was in his cabin at the time of the collision. The CRYSTAL's bow directly struck his cabin, located above the waterline. The impact severely damaged his cabin, trapping him inside. The CO called the bridge requesting assistance.

17

Five Sailors used a sledgehammer, kettlebell, and their bodies to break through the door into the CO's cabin, remove the hinges, and then pry the door open enough to squeeze through. Even after the door was open, there was a large amount of debris and furniture against the door, preventing anyone from entering or exiting easily.

A junior officer and two chief petty officers removed debris from in front of the door and crawled into the cabin. The skin of the ship and outer bulkhead were gone and the night sky could be seen through the hanging wires and ripped steel. The rescue team tied themselves together with a belt in order to create a makeshift harness as they retrieved the CO, who was hanging from the side of the ship.

The team took the CO to the bridge, where a medical team assessed his condition. As he was being monitored by personnel on the bridge, his condition worsened. A team of stretcher bearers moved the CO from the bridge to the at-sea cabin at 0319, and shortly thereafter, due to the severity of his injuries, he was medically evacuated from the ship at 0710 to USNHY via helicopter. He was treated and released on 18 June 2017.

4. SEARCH AND RESCUE AND ASSISTANCE

The Japanese Coast Guard sent the vessels IZANAMI and KANO from Shimoda to assist FITZGERALD. IZANAMI arrived at approximately 0452. Japanese helicopters remained in the vicinity of FITZGERALD and assisted in search and rescue efforts, along with U.S. Maritime Patrol aircraft.

A Japanese Coast Guard helicopter arrived to medically evacuate the Commanding Officer by lowering a rescue litter (basket) onto the deck while hovering above FITZGERALD because the significant list of the ship to the starboard side prevented the helicopter from landing. The Japanese Coast Guard helicopter then transported the CO to the Navy Hospital in Yokosuka.

At 0745, USS DEWEY (DDG 105) executed an emergency underway from Yokosuka Naval Station to assist FITZGERALD.

At 0914 all water levels were reported as holding steady. At 0911, Deputy Commodore, Destroyer Squadron FIFTEEN arrived via helicopter, along with a doctor, DCC, and Chaplain to support the crew. One injured Officer and one Sailor who escaped Berthing 2 were loaded onto the aircraft using a litter and were medically evacuated on the same aircraft as it departed. These Sailors were brought to the Navy Hospital in Yokosuka.

At approximately 1000, FITZGERALD requested a Rescue and Assistance Team and additional DC pumps and hoses from DEWEY. At approximately 1200, the Dewey team and 14 additional personnel with an additional P-100 pump and firehoses embarked FITZGERALD to assist. Between the hours of 1200 and 1900 the DEWEY team assisted with dewatering, provided food and water, and remained aboard FITZGERALD until arrival pierside in Yokosuka.

At 1226, two Japanese Self Defense Force (JMSDF) Helicopters, one JMSDF P-3C, one

USN P-8, and one USN helicopter were in the area of the collision helping to support the search and rescue effort.

5. TRANSIT TO YOKOSUKA

FITZGERALD was underway at 0453 at 3 knots. The ship was able to begin the transit to homeport under her own power. Because FITZGERALD would take on additional water if the ship moved too quickly on the transit home, speed was limited between 3 and 5 knots. Once flooding was stabilized, the list was held at 5 degrees to the right (starboard side).

At 0815, two tugboats dispatched from Port Operations in Yokosuka Naval Base came alongside FITZGERALD. The tugs were driven by U.S. Navy Harbor pilots and approached carefully due to FITZGERALD's severe list. One tugboat was stationed in front of FITZGERALD with a line to FITZGERALD's bow to help tow. The second tugboat was configured in a "power make up" approximately half-way down the length of FITZGERALD on the left (port) side. This configuration allowed the tug to maintain control of FITZGERALD in case the ship lost all propulsion and steering and could not move under own power.

Steering in a straight line was challenging given the damage FITZGERALD had sustained. The ship lost the ability to steer from the normal location in the pilothouse. Steering was conducted from Aft Steering control station, an arrangement not often used but frequently practiced.

At 1607, FITZGERALD entered the inner harbor of Yokosuka Naval Base. In order to set up an effective separation scheme (similar to traffic lanes), Yokosuka Naval Base Port Operations hired a commercial tugboat to escort and provide an additional buffer against other harbor traffic. At 1854 FITZGERALD moored pier-side in Yokosuka. The tugboats maneuvered to minimize water from the tugboats going into the ship, and assisted in mooring FITZGERALD to the pier.

6. DIVING AND RECOVERY OPERATIONS

Once FITZGERALD was back in Yokosuka, a Navy dive team conducted two dives. The first dive occurred on the evening of 17 June 2017, and focused on doing an initial assessment of the ship and looking for ways for divers to enter into the damaged spaces to search for missing personnel.

The second dive began at 0454 the following morning, 18 June 2017. The divers entered the Berthing 2 space through the hole in the starboard side of the ship. The divers immediately found GMSN Dakota Rigsby near the top of the starboard access trunk. His foot was caught between the ladder and the wall but, as the foot was easily released by the divers, there was no indication of whether this happened while he was trying to leave or if he floated to this position after he passed away. GMSN Rigsby's body was brought to the dive boat at 0523.

The divers went back into the water at 0611 and entered the ship. Once back inside the Berthing 2 space, they immediately found YN2 Shingo Alexander Douglass, FC1 Carlos

Sibayan, PSC Xavier Martin and STG2 Ngoc Truong Huynh. These Sailors were found in the lounge area of Berthing 2. STG2 Ngoc Truong Huynh's body was found underneath a television, but it did not appear that he had been pinned by the television. GM1 Hernandez was found in the main passageway of Berthing 2 nearest the lounge area. Along with GMSN Rigsby, these Sailors were all found on the starboard side of Berthing 2.

The door to the bathroom in Berthing 2 was closed. When the divers entered the bathroom, they found FCC Gary Rehm just inside this space.

7. INJURIES

The Sailor who escaped the starboard side of Berthing 2 suffered near drowning, seawater aspiration, traumatic brain injury, scalp laceration, and ankle laceration. He was medically evacuated to the USNHY because of the severity of his injuries and released on 19 June 2017.

The Combat Systems Officer suffered traumatic brain injury with brief loss of consciousness and facial abrasions and contusions. He was medically evacuated to the U.S. Navy Hospital in Yokosuka because of the severity of his injuries and released on 18 June 2017.

The CO suffered traumatic brain injury with brief loss of consciousness, back contusion and lower extremity abrasions. He was medially evacuated to the U.S. Navy hospital in Yokosuka because of the severity of his injuries and released on 18 June 2017.

Seven Sailors were unable to egress the space and died. The loss of seven shipmates is a tragedy beyond words and a reminder of the dangers inherent in the mission of every ship and Sailor.

8. FINDINGS

Collisions at sea are rare and the relative performance and fault of the vessels involved is an open admiralty law issue. The Navy is not concerned about the mistakes made by CRYSTAL. Instead, the Navy is focused on the performance of its ships and what we could have done differently to avoid these mishaps.

In the Navy, the responsibility of the Commanding Officer for his or her ship is absolute. Many of the decisions made that led to this incident were the result of poor judgment and decision making of the Commanding Officer. That said, no single person bears full responsibility for this incident. The crew was unprepared for the situation in which they found themselves through a lack of preparation, ineffective command and control, and deficiencies in training and preparations for navigation.

8.1 Training

FITZGERALD officers possessed an unsatisfactory level of knowledge of the International Rules of the Nautical Road.

Watch team members were not familiar with basic radar fundamentals, impeding effective use.

8.2 Seamanship and Navigation

The Officer of the Deck and bridge team failed to comply with the International Rules of the Nautical Road. Specifically:

FITZGERALD was not operated at a safe speed appropriate to the number of other ships in the immediate vicinity.

FITZGERALD failed to maneuver early as required with risk of collision present.

FITZGERALD failed to notify other ships of danger and to take proper action in extremis.

Watch team members responsible for radar operations failed to properly tune and adjust radars to maintain an accurate picture of other ships in the area.

Watchstanders performing physical look out duties did so only on FITZGERALD's left (port) side, not on the right (starboard) side where the three ships were present with risk of collision.

Key supervisors responsible for maintaining the navigation track and position of other ships:

Were unaware of existing traffic separation schemes and the expected flow of traffic.

Did not utilize the Automated Identification System. This system provides real time updates of commercial ship positions through use of the Global Positioning System.

FITZGERALD's approved navigation track did not account for, nor follow, the Vessel Traffic Separation Schemes in the area.

8.3 Leadership and Culture

The bridge team and Combat Information Center teams did not communicate effectively or share information. The Combat Information Center is the space on U.S. Surface Ships where equipment and personnel combine to produce the most accurate picture of the operating environment.

Supervisors and watch team members on the bridge did not communicate information and concerns to one another as the situation developed.

The Officer of the Deck, responsible for the safe navigation of the ship, did not call the Commanding Officer on multiple occasions when required by Navy procedures.

Key supervisors in the Combat Information Center failed to comprehend the complexity of the operating environment and the number of commercial vessels in the area.

In several instances, individual members of the watch teams identified incorrect information or mistakes by others, yet failed to proactively and forcefully take corrective action, or otherwise highlight or communicate their individual concerns.

Key supervisors and operators accepted difficulties in operating radar equipment due to material faults as routine rather than pursuing solutions to fix them.

The command leadership did not foster a culture of critical self-assessment. Following a near-collision in mid-May, leadership made no effort to determine the root causes and take corrective actions in order to improve the ship's performance.

The command leadership was not aware that the ship's daily standards of performance had degraded to an unacceptable level.

8.4 Fatigue

The command leadership allowed the schedule of events preceding the collision to fatigue the crew.

The command leadership failed to assess the risks of fatigue and implement mitigation measures to ensure adequate crew rest.

This assessment of USS FITZGERALD is not intended to imply that CRYSTAL mistakes and deficiencies were not also factors in the collision.

ANNEX A – TIMELINE OF EVENTS

<u>16 June 2017</u>

0001	FITZGERALD is moored at Commander, Fleet Activities Yokosuka (CFAY) Pier 12.
0600	Liberty expires for crew.
0900	Navigation briefing held to prepare crew for underway and anchoring evolution.
1030	Stationed "Sea and Anchor" Detail to get underway.
1130	Underway from Pier 12.
1210	Anchored in preparation for ammunition on-load.
1545	Stationed the Sea and Anchor Detail to get underway from anchor. The Sea and Anchor Detail provides additional personnel experienced in navigating in restricted waters.
1624	Underway from anchor.
1736	Conducted helicopter deck landing qualifications and aviation certifications.
1747	Stationed Modified Navigation Detail due to proximity of shoal water while transition to Sagami Wan. The Modified Navigation detail provides additional personnel when navigating in close proximity to shallow water.
1835	The Modified Navigation Detail was secured.
1859	Sunset was observed and navigation lights were energized and dimmed to support flight operations.
2111	Flight operations were secured.
2116	The Modified Navigation Detail was stationed while closing land in preparation for small boat operations to return Afloat Training Group Western Pacific personnel ashore.
By 2200	All watchstanders assigned to the 2200-0200 watch time period were on duty.

Approx 2300	Small Boat Operations secured and FITZGERALD proceeded southwest from Sagami Wan to sea.
2300	The Commanding Officer left the Bridge.
Approx 2305	The Executive Officer and Navigator left the Bridge.
2311-2345	FITZGERALD maintained 16 knots due to high traffic density.
2330	Moonrise at 69% illumination.
2345	FITZGERALD maneuvered to course 230 and increased speed to 20 knots
2350	FITZGERALD overtook a contact on the left (port) side within 3 nautical miles and no report was made to the Commanding Officer as required by his Standing Orders procedures. No course and speed determinations were made for this vessel by watchstanders.

17 June 2017

0000	FITZGERALD approached the Vessel Traffic Separation Scheme (VTSS) north of Oshima Island.
0000	FITZGERALD was in vicinity of four commercial vessels, two of which were within 3 nautical miles and no report was made to the Commanding Officer as required by his Standing Orders procedures. No course and speed determinations were made for this vessel by watchstanders.
0015	FITZGERALD was passing two commercial vessels, one of which was within 3 nautical miles and no report was made to the Commanding Officer as required by his Standing Orders procedures. No course and speed determinations were made for this vessel by watchstanders.
0022	FITZGERALD altered course to 220 and remained at 20 knots.
0033	FITZGERALD altered course to 215.
0034	Four vessels passed down the left (port) side with closest point of approach at 1500 yards. The Commanding Officer was informed. No course and speed determinations were made for these vessels. Radar contact on them was not held.
0054	FITZGERALD altered course to 190 while remaining at 20 knots.
0058	FITZGERALD was in the vicinity of five commercial vessels. Three of these passed on the left (port) side within 3 nautical miles and no report was made to the Commanding Officer as required by his Standing Orders

procedures. No course and speed determinations were made for this vessel by watchstanders.

0100	FITZGERALD remained on course 190 at 20 knots.

0108	FITZGERALD crossed the bow of a ship at approximately 650 yards, passed a second vessel at 2 nautical miles, and a third vessel at 2.5 nautical miles. No reports were made to the Commanding Officer as required by his Standing Orders procedures. No course and speed determinations were made for this vessel by watchstanders.

0110	FITZGERALD continued on course 190, speed 20 knots. CRYSTAL was ahead on FITZGERALD's starboard side at a distance of 11 nautical miles.

0110	Watchstanders unsuccessfully attempted to initiate a radar track on the CRYSTAL.

0115	CRYSTAL was closing FITZGERALD's intended track at a high rate of speed.

0117	The FITZGERALD Officer of the Deck plotted a radar track on a vessel thought to be CRYSTAL and calculated that CRYSTAL would pass 1500 yards from FITZGERALD on the right (starboard) side. It is unknown if the OOD was tracking the CRYSTAL or another commercial vessel.

0120	The watch stander responsible for immediate support to the Officer of the Deck, the Junior Officer of the Deck, reported sighting CRYSTAL visually and noted that CRYSTAL's course would cross FITZGERALD's track. The Officer of the Deck continued to think that CRYSTAL would pass at 1500 yards from FITZGERALD.

0122	The Junior Officer of the Deck sighted CRYSTAL again and made the recommendation to slow. The Officer of the Deck responded that slowing would complicate the contact picture.

0125	CRYSTAL was approaching FITZGERALD from the right (starboard) side at 3 nautical miles. FITZGERALD watchstanders at this time held two other commercial vessels in addition to CRYSTAL. One was calculated to have closest approach point at 2000 yards and the other was calculated to risk collision. No contact reports were made to the Commanding Officer and no additional course and speed determinations were made on these vessels.

25

0125	The Officer of the Deck noticed CRYSTAL rapidly getting closer and considered a turn to 240T.
0127	The Officer of the Deck ordered course to the right to course 240T, but rescinded the order within a minute. Instead, the Officer of the Deck ordered an increase to full speed and a rapid turn to the left (port). These orders were not carried out.
0129	The Bosun Mate of the Watch, a more senior supervisor on the bridge, took over the helm and executed the orders.
As of 0130	Neither FITZGERALD nor CRYSTAL made an attempt to establish radio communications or sound the danger signal.
As of 0130	FITZGERALD had not sounded the collision alarm.
0130:34	CRYSTAL's bow struck FITZGERALD at approximately frame 160 on the right (starboard) side above the waterline and CRYSTAL's bulbous bow struck at approximately frame 138 below the waterline.

ANNEX B – PHOTOGRAPHS AND DRAWINGS

B1: Commanding Officer's Stateroom on FTZ - Exterior View

TOP

B2: Commanding Officer's Stateroom on FTZ - Interior view

TOP

B3: Starboard Access Trunk outside Berthing 2 (FTZ and BEN)

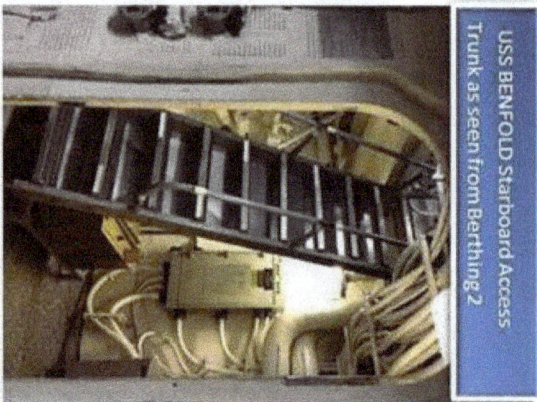

USS BENFOLD Starboard Access Trunk as seen from Berthing 2

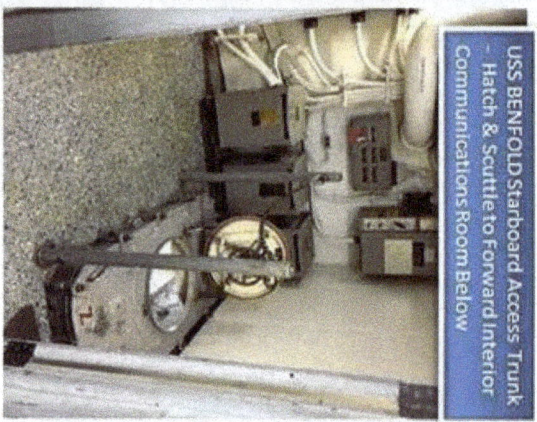

USS FITZGERALD Starboard Access Trunk as seen from Berthing 2

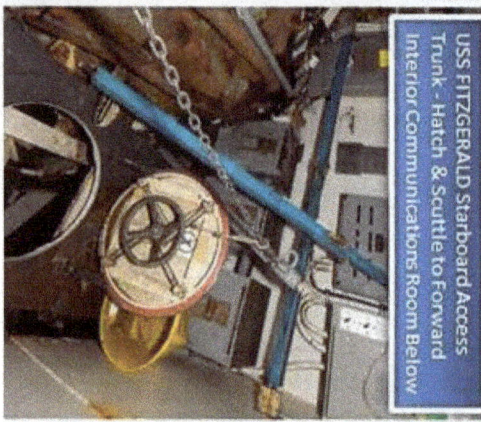

USS BENFOLD Starboard Access Trunk – Hatch & Scuttle to Forward Interior Communications Room Below

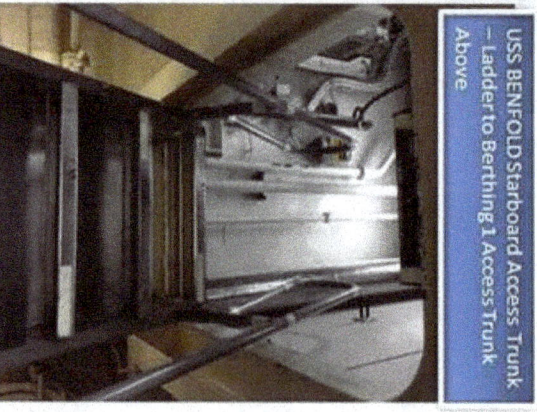

USS FITZGERALD Starboard Access Trunk - Hatch & Scuttle to Forward Interior Communications Room Below

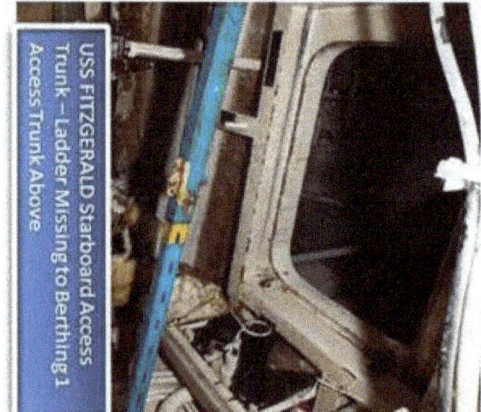

USS BENFOLD Starboard Access Trunk – Ladder to Berthing 1 Access Trunk Above

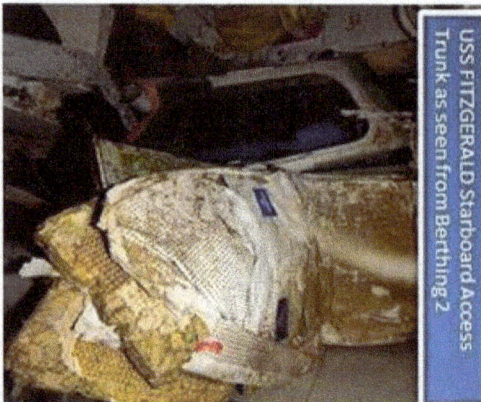

USS FITZGERALD Starboard Access Trunk – Ladder Missing to Berthing 1 Access Trunk Above

B6: Lounge in Berthing 2 - View from racks

TOP

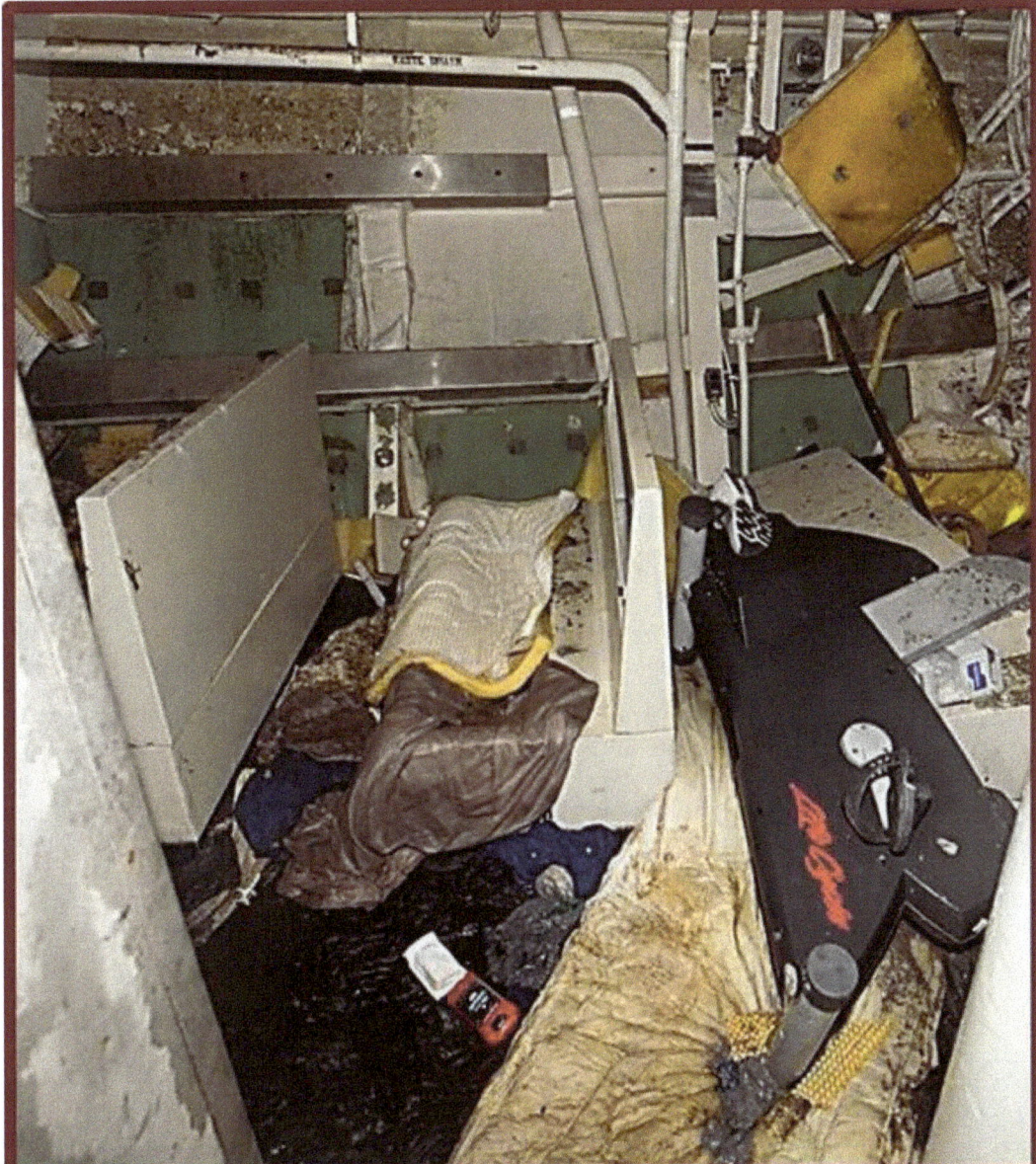

B7: Sailor sketch depicting chest high water from top rack

B8: Sailor sketch depicting egress from starboard section of Berthing 2 (above water)

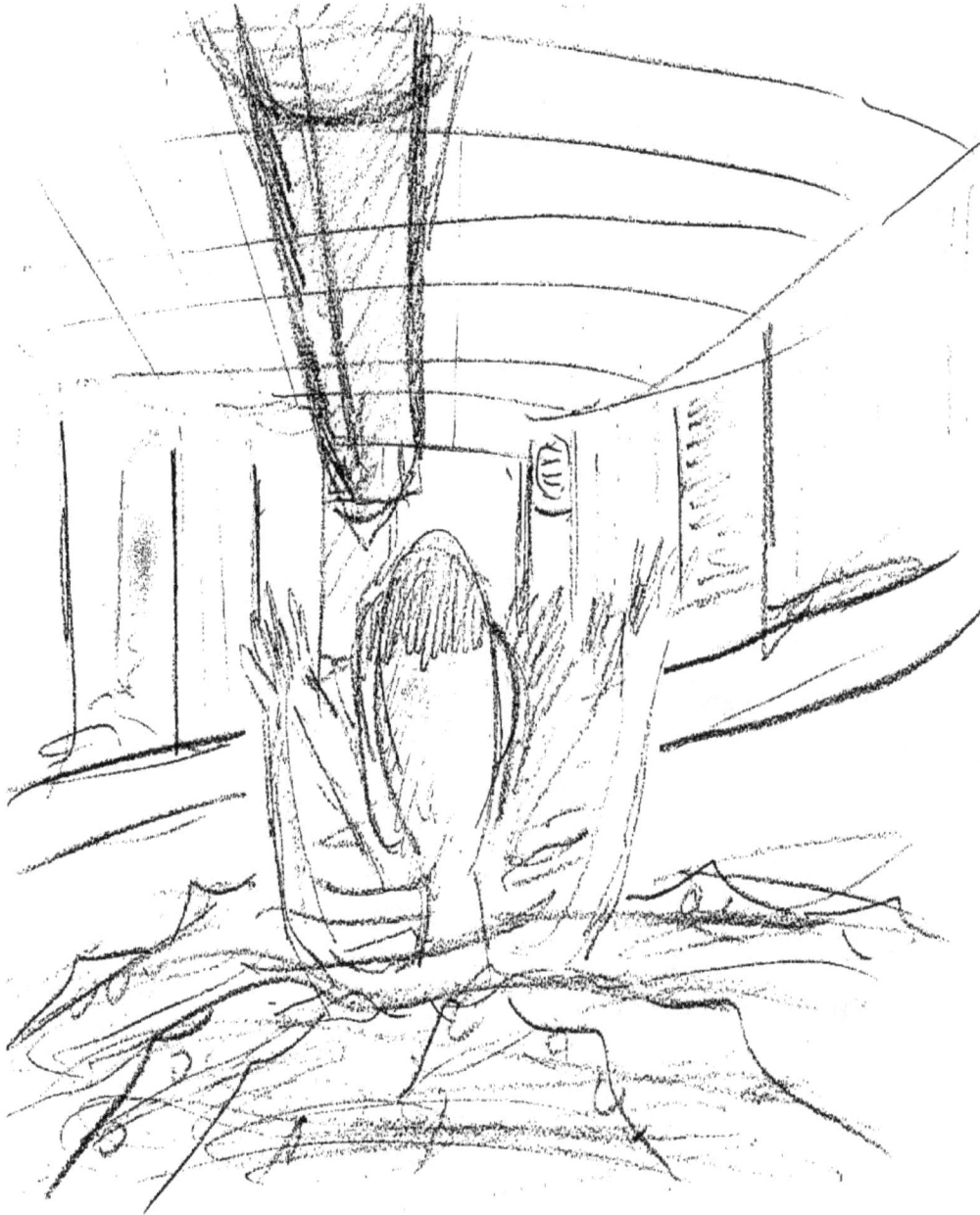

B9: Sailor sketch depicting egress from starboard section of Berthing 2 (below water)

B10: Sailor sketch depicting ceiling of Berthing 2

SELF POINT OF VIEW IN 2ᴺᴰ AIR POCKET

B11: Sailor sketch depicting path of starboard egress from Berthing 2

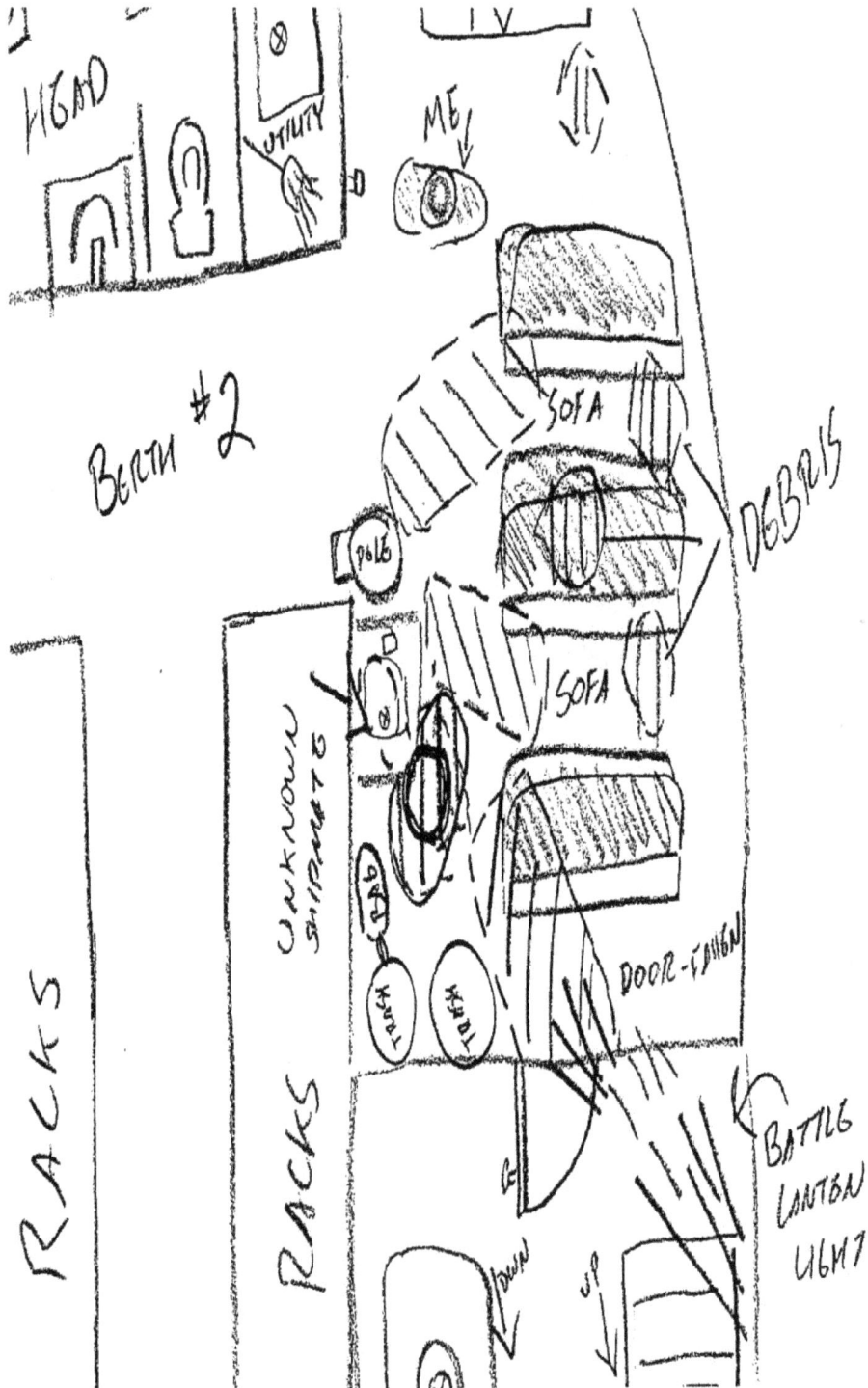

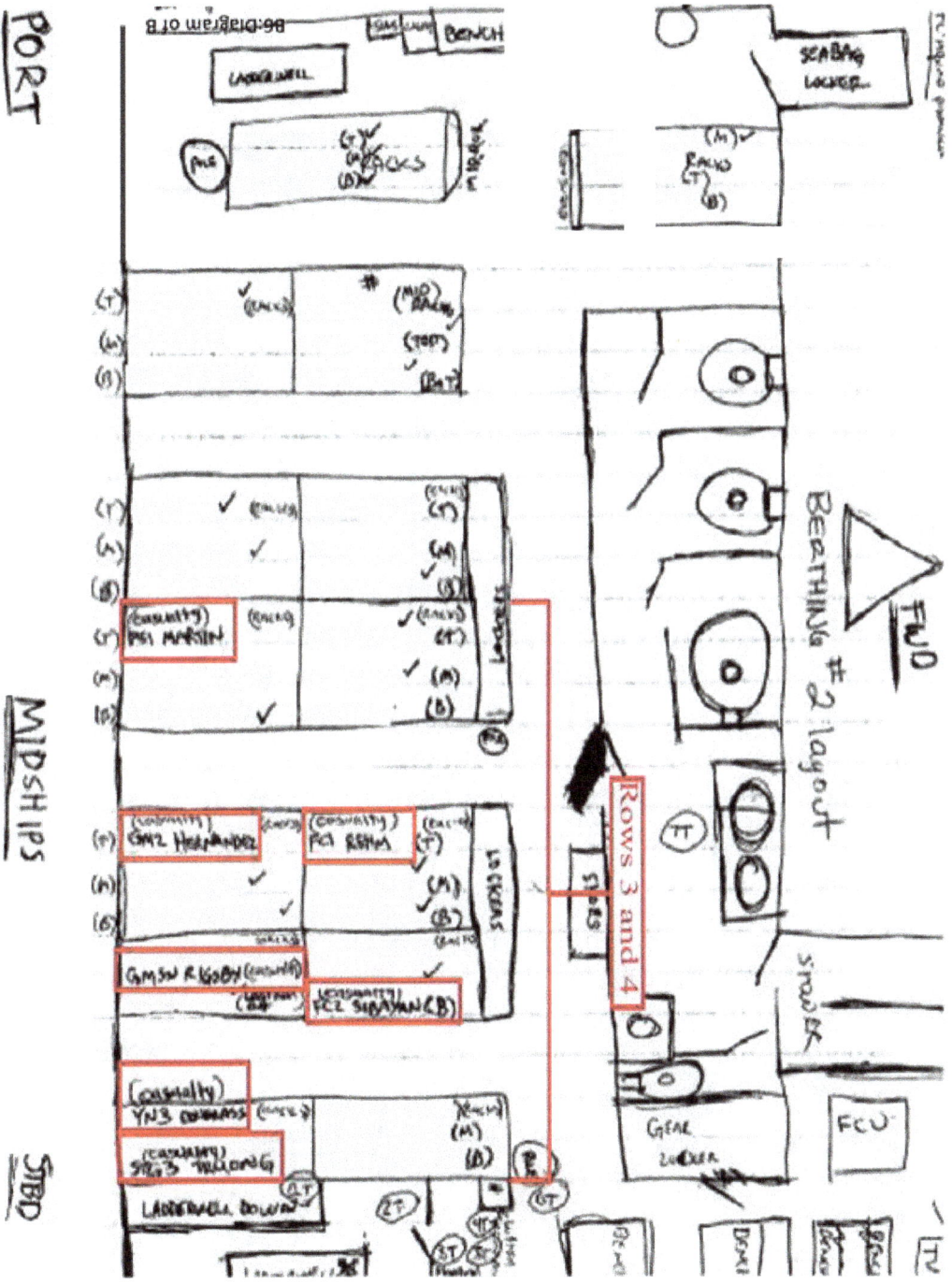

B13: Diagram of Berthing 2 (annotated with Tango recoveries)

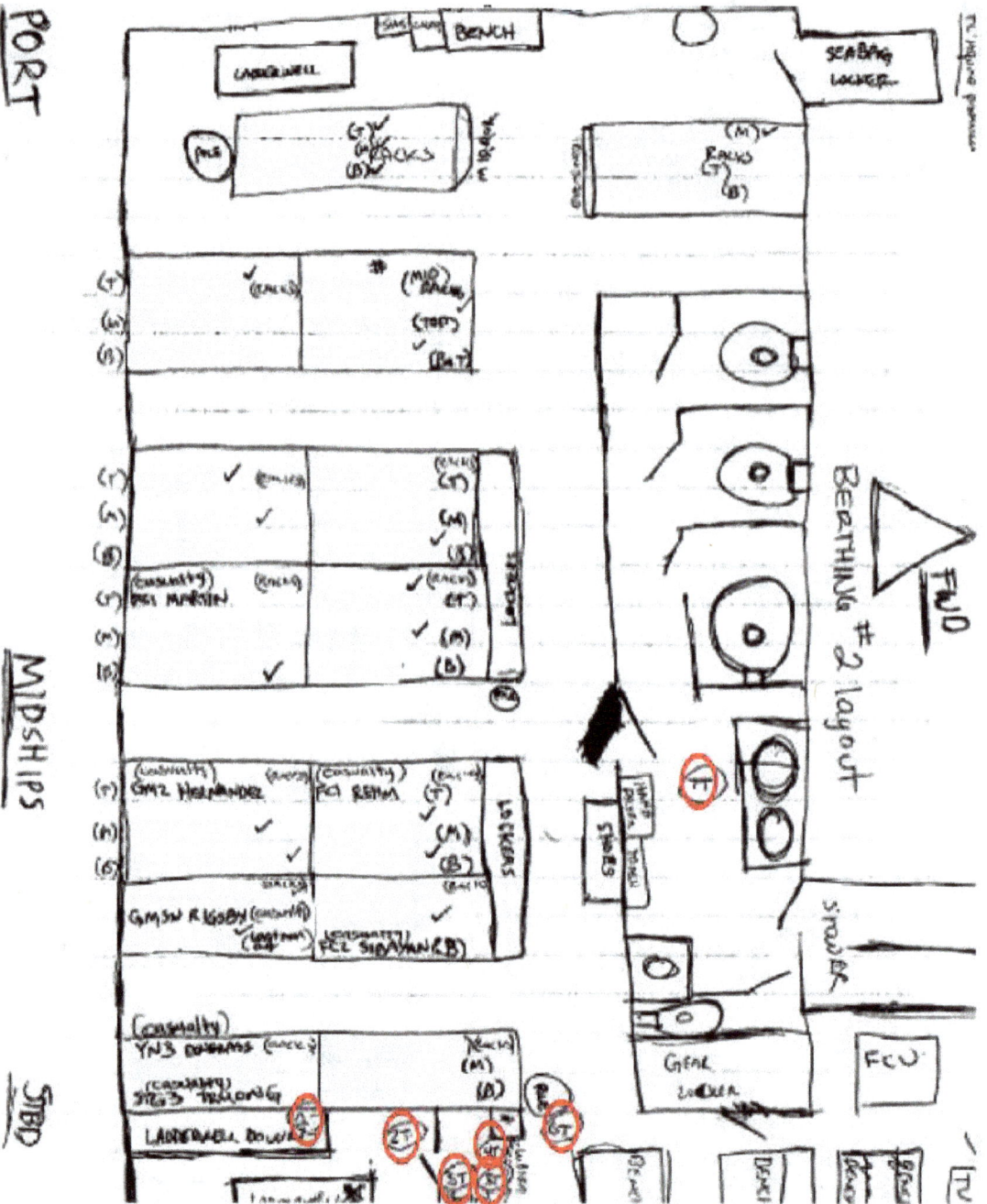

ANNEX C –Summary of Flooding and Structural Damage to USS FITZGERALD

1.   The following spaces were flooded due to the hole in the side of the ship and proximity flooding, which is flooding into spaces directly connected to areas near the hole and which are not separated by a water-tight barrier. This means water flowed from the hole in the ship into these spaces through doors and fixtures that are not designed to be water-tight. Flooded spaces marked with (**) also suffered structural damage.

       a.   Berthing 2 (3-97-02-L)
       b.   Berthing 2 Starboard Vestibule (Access Trunk) (3-116-1-T) **
       c.   Berthing 2 Seabag Locker (3-78-1-A)
       d.   Berthing 2 Seabag Locker (3-78-2-A)
       e.   Berthing 2 Head (3-97-01-L)
       f.   Berthing 2 Cleaning Gear Locker (3-102-1-A)
       g.   Auxiliary Machinery Room Number 1 (4-126-0-E) **
       h.   Forward Interior Communications Room (4-94-0-C)
       i.   Forward Vacuum Collection Holding Tank Room (4-110-0-E)
       j.   Forward Vacuum Collection Holding Tank Room Passageway (4-110-1-L)
       k.   Chemical, Biological Radiological Clothing Store Room (4-110-3-A)
       l.   Berthing 1 (2-78-01-L)
       m.   Berthing 1 Port Access Trunk (2-116-2-T)
       n.   Berthing 1 Head (2-97-01-L)
       o.   Berthing 1 Cleaning Gear Locker (2-102-1-A)
       p.   Berthing 2 / Forward Interior Communications Room Access Trunk (3-97-2-T)
       q.   Auxiliary Machinery Room Number 1 Escape Trunk (3-97-1-T)
       r.   Auxiliary Machinery Room Number 1 Port Escape Trunk (4-122-2-T)

2.   The following spaces were flooded due to cross flooding, which is the flooding of spaces that are connected to damaged spaces and that have the ability to be isolated with a water tight barrier, but which could not be sealed off in time. Flooded spaces marked with (**) also suffered structural damage.

       a.   Auxiliary Machinery Room Number 1 Airlock (3-158-1-L)
       b.   Auxiliary Machinery Room Number 1 Vestibule (3-158-3-T)
       c.   Sonar Control Room (2-50-2-C)
       d.   Sonar Control Fan Room (2-42-2-Q)
       e.   Sonar Control Passageway (2-46-0-L)
       f.   Sonar Control Admin Office (2-42-1-Q)
       g.   Combat Systems Equipment Room 1(2-53-1-C)
       h.   Forward 400Hz Power Conversion Room (3-126-2-Q)
       i.   Forward 400Hz Fan Room (3-164-2-Q)
       j.   Radio Vestibule (Access Trunk) (2-158-3-T)
       k.   Radio Central (2-126-1-C)
       l.   Radio Transmitter Room (2-158-1-C)
       m.   Main Engine Room Number 1 (4-174-0-E)

3.   The following spaces were partially flooded due to ruptured firemain (large seawater pipes that provide water for fighting fires) and ruptured AFFF (Aqueous Film-Forming Foam) tank. Flooded spaces marked with (**) also suffered structural damage.

    a.   Combat Information Center Passageway (1-126-3-L) **
    b.   Repair Locker Number 2 Passageway (1-78-01-L) **
    c.   5" Projectile Magazine (3-42-0-M)
    d.   5" Powder Magazine Number 1 (3-68-2-M)
    e.   5" Powder Magazine Number 2 (3-62-1-M)
    f.   5" Magazine Access Trunk (3-42-1-T)
    g.   Combat Information Center Starboard Access Passageway (1-158-1-L)
    h.   Forward Pallet Staging Area (1-42-01-L)
    i.   Sonar Cooling Equipment Room (4-42-0-Q)
    j.   Crew Messline Passageway (1-174-01-L)
    k.   Operations Office (1-84-1-Q)
    l.   Electronic Key Management System (EKMS) Vault (1-110-3-Q)
    m.   Sonar Room Number 1 (4-42-0-Q)

4.   The following spaces suffered structural damage.

    a.   Commanding Officer Cabin (02-146-1-L)
    b.   Commanding Officer Stateroom (02-136-3-L)
    c.   Commanding Officer Bath (02-126-3-L)
    d.   Stateroom (02-158-7-L)
    e.   SPY Radar Array Room Number 1 (03-126-1-Q)
    f.   Electronic Workshop Number 1 (03-142-1-Q) g.   Fan Room (03-142-3-Q)
    h.   SPY Radar Array Room Number 3 (03-158-3-Q)
    i.   Radar Room Number 1 (03-128-0-C) j.   Fan Room (01-126-3-Q),
    k.   Combat Systems Maintenance Central Airlock (01-126-1-L)
    l.   Starboard Break (01-118-1-L),
    m.   Combat Systems Maintenance Central Ladder well (01-110-1-L)

ENCLOSURE (2)

REPORT ON THE COLLISION BETWEEN USS JOHN S MCCAIN (DDG 56)
AND MOTOR VESSEL ALNIC MC

23 OCT 2017

42

1. EXECUTIVE SUMMARY - USS JOHN S MCCAIN

1.1 Introduction

USS JOHN S MCCAIN collided with Motor Vessel ALNIC MC on 21 August 2017 in the Straits of Singapore.

JOHN S MCAIN is a Flight 1 Arleigh Burke Class Destroyer, commissioned in 1994 and homeported in Yokosuka, Japan, as part of the Forward Deployed Naval Forces and Carrier Strike Group FIVE. Approximately 300 sailors serve aboard MCCAIN. MCCAIN is 505 feet in length and carries a gross tonnage of approximately 9,000 tons.

ALNIC MC is a Liberia flagged oil and chemical tanker built in 2008. ALNIC MC is approximately 600 feet long and has a gross tonnage of approximately 30,000 tons.

The collision between JOHN S MCCAIN and ALNIC resulted in the deaths of 10 U.S. Sailors due to impact with MCCAIN's berthing compartments, located below the waterline of the ship. ALNIC suffered no fatalities. U.S. Sailor fatalities were:

ETC Charles Nathan Findley of Amazonian, Missouri, 31 years old.

ICC Abraham Lopez of El Paso, Texas, 39 years old.

ET1 Kevin Sayer Bushell of Gaithersburg, Maryland, 26 years old.

ET1 Jacob Daniel Drake of Cable, Ohio, 21 years old.

IT1 Timothy Thomas Eckels Jr. of Baltimore, Maryland, 23 years old.

IT1 Corey George Ingram of Poughkeepsie, New York, 28 years old.

ET2 Dustin Louis Doyon of Suffield, Connecticut, 26 years old.

ET2 John Henry Hoagland III of Killeen, Texas, 20 years old.

IC2 Logan Stephen Palmer of Harristown, Illinois, 23 years old.

ET2 Kenneth Aaron Smith of Cherry Hill, New Jersey, 22 years old.

1.2 Summary of Findings

The Navy determined the following causes of the collision:

Loss of situational awareness in response to mistakes in the operation of the JOHN S MCCAIN's steering and propulsion system, while in the presence of a high density of maritime traffic.

Failure to follow the International Nautical Rules of the Road, a system of rules to govern the maneuvering of vessels when risk of collision is present.

Watchstanders operating the JOHN S MCCAIN's steering and propulsion systems had insufficient proficiency and knowledge of the systems.

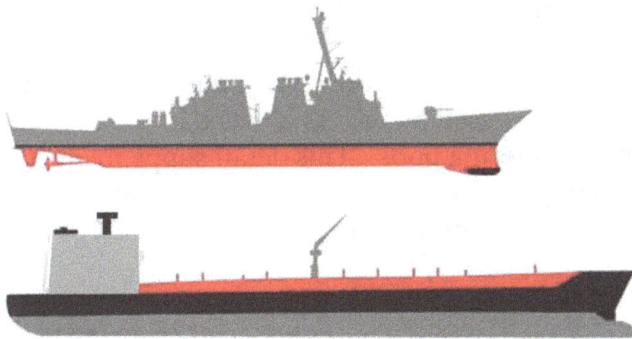

Figure 1 – Relative size of USS JOHN S MCCAIN

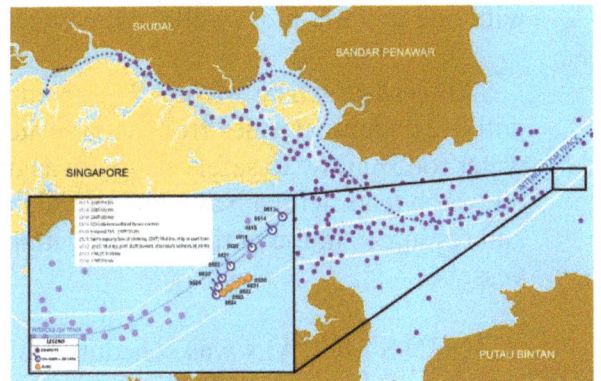

Figure 2 – Illustration Map of Approximate Collision Location

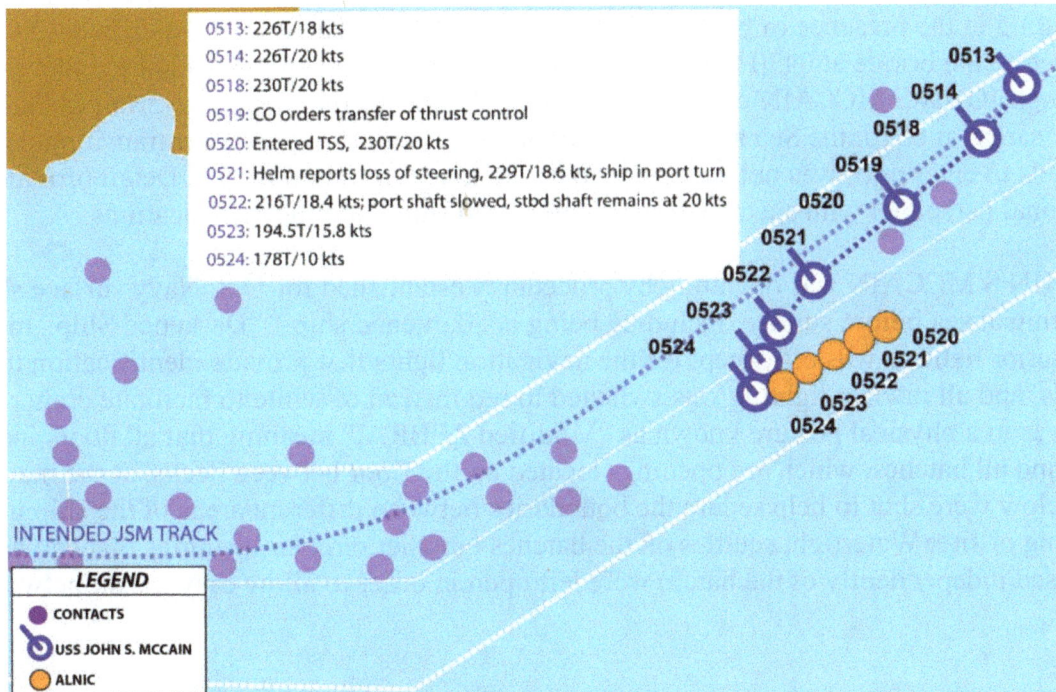

0513: 226T/18 kts
0514: 226T/20 kts
0518: 230T/20 kts
0519: CO orders transfer of thrust control
0520: Entered TSS, 230T/20 kts
0521: Helm reports loss of steering, 229T/18.6 kts, ship in port turn
0522: 216T/18.4 kts; port shaft slowed, stbd shaft remains at 20 kts
0523: 194.5T/15.8 kts
0524: 178T/10 kts

INTENDED JSM TRACK

LEGEND
● CONTACTS
◉ USS JOHN S. MCCAIN
● ALNIC

Figure 3 – Illustration Map of Approximate Collision Location

44

2. DESCRIPTION OF EVENTS

2.1 Background

JOHN S MCCAIN departed its homeport of Yokosuka, Japan on 26 May 2017 for a scheduled six month deployment in the Western Pacific, which at the time of the collision had included operations in the East and South China Seas, and port visits in Vietnam, Australia, Philippines and Japan. On the morning of 21 August, JOHN S MCCAIN was 50 nautical miles east of Singapore, approaching the Singapore Strait and Strait of Malacca, in transit to a scheduled port of call at Changi Naval Base, Singapore. These Straits form a combined ocean passage that is one of the busiest shipping lanes in the world, with more than 200 vessels passing through the straits each day. JOHN S MCCAIN was transiting through the southern end of the Strait. See Figure 2. In the predawn hours of 21 August 2017, the moon had set and the skies were overcast. There was no illumination and the sun would not rise until 0658. Seas were calm, with one to three foot swells. All navigation and propulsion equipment was operating properly.

At 0418, JOHN S MCCAIN transitioned to a Modified Navigation Detail due to approaching within 10 nautical miles from shoal water. This detail is used by the Navy when in proximity of water too shallow to safely navigate as occurs when entering ports. This detail supplemented the on watch team with a Navigation Evaluator and Shipping Officer, providing additional personnel and resources in the duties of Navigation and management of the ship's relative position to other vessels.

JOHN S MCCAIN was scheduled to enter the Singapore Strait Traffic Separation Scheme less than an hour later. Traffic separation schemes are established by local authorities in approaches to ports throughout the world to provide ships assistance in separating their movements when transiting to and from ports. The Commanding Officer had been physically present on the bridge since 0115, a practice common for operations with higher risk, such as navigating in the presence of busy maritime traffic at night. The Executive Officer (XO) reported to the bridge at 0430 to provide additional supervision and oversight to enter port. Although JOHN S MCCAIN entered the Middle Channel of the Singapore Strait (a high traffic density area) at 0520, the Sea and Anchor Detail, a team the Navy uses for transiting narrower channels to enter port, was not scheduled to be stationed until 0600. This Detail provides additional personnel with specialized navigation and ship handling qualifications.

JOHN S MCCAIN was operating by procedures established for U.S. Navy surface ships when operating at sea before sunrise, including being at "darkened ship." "Darkened Ship" means that all exterior lighting was off except for the navigation lights that provide identification to other vessels, and all interior lighting was switched to red instead of white to facilitate crew rest. The ship was in a physical posture known as "Modified ZEBRA," meaning that all doors inside the ship, and all hatches, which are openings located on the floor between decks, at the main deck and below were shut to help secure the boundaries between different areas of the ship in case of flooding or fire. Watertight scuttles on the hatches (smaller circular openings that can be opened or closed independently of the hatch) were left open in order to allow easy transit between spaces.

2.2 Events Leading to the Collision

At 0519, the Commanding Officer noticed the Helmsman (the watchstander steering the ship) having difficulty maintaining course while also adjusting the throttles for speed control. In response, he ordered the watch team to divide the duties of steering and throttles, maintaining course control with the Helmsman while shifting speed control to another watchstander known as the Lee Helm station, who sat directly next to the Helmsman at the panel to control these two functions, known as the Ship's Control Console. See Figures 3 and 4. This unplanned shift caused confusion in the watch team, and inadvertently led to steering control transferring to the Lee Helm Station without the knowledge of the watch team. The CO had only ordered speed control shifted. Because he did not know that steering had been transferred to the Lee Helm, the Helmsman perceived a loss of steering.

Figure 4 – Bridge Schematic of JOHN S MCCAIN

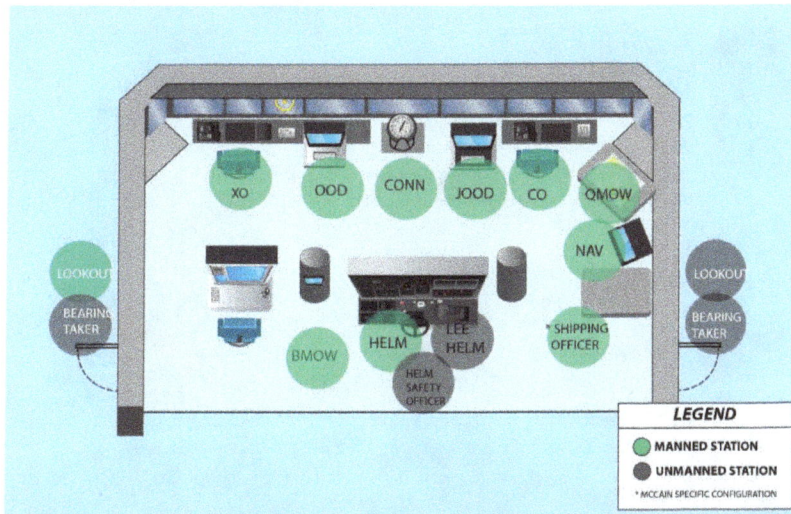

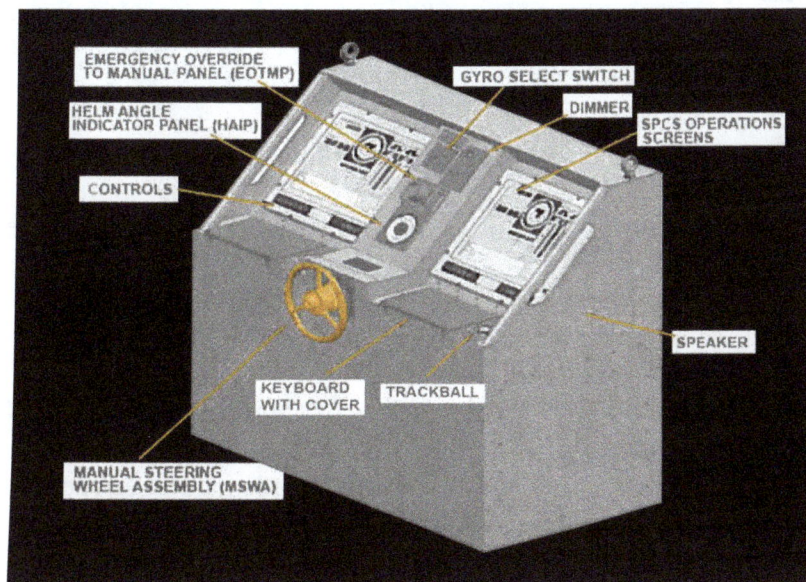

Figure 5 – Illustration of Ship Control Console on JOHN S MCCAIN

46

Steering was never physically lost. Rather, it had been shifted to a different control station and watchstanders failed to recognize this configuration. Complicating this, the steering control transfer to the Lee Helm caused the rudder to go amidships (centerline). Since the Helmsman had been steering 1-4 degrees of right rudder to maintain course before the transfer, the amidships rudder deviated the ship's course to the left.

Additionally, when the Helmsman reported loss of steering, the Commanding Officer slowed the ship to 10 knots and eventually to 5 knots, but the Lee Helmsman reduced only the speed of the port shaft as the throttles were not coupled together (ganged). The starboard shaft continued at 20 knots for another 68 seconds before the Lee Helmsman reduced its speed. The combination of the wrong rudder direction, and the two shafts working opposite to one another in this fashion caused an un-commanded turn to the left (port) into the heavily congested traffic area in close proximity to three ships, including the ALNIC. See Figure 5.

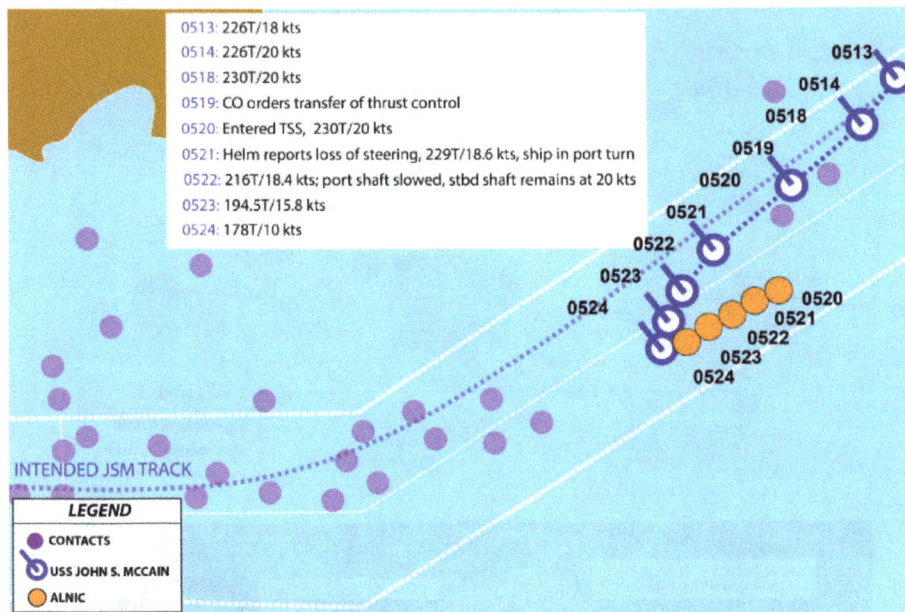

Figure 6 – Illustration Map of Approximate Collision Location

Although JOHN S MCCAIN was now on a course to collide with ALNIC, the Commanding Officer and others on the ship's bridge lost situational awareness. No one on the bridge clearly understood the forces acting on the ship, nor did they understand the ALNIC's course and speed relative to JOHN S MCCAIN during the confusion.

Approximately three minutes after the reported loss of steering, JOHN S MCCAIN regained positive steering control at another control station, known as Aft Steering, and the Lee Helm gained control of both throttles for speed and corrected the mismatch between the port and starboard shafts. These actions were too late, and at approximately 0524 JOHN S MCCAIN crossed in front of ALNIC's bow and collided. See Figure 6.

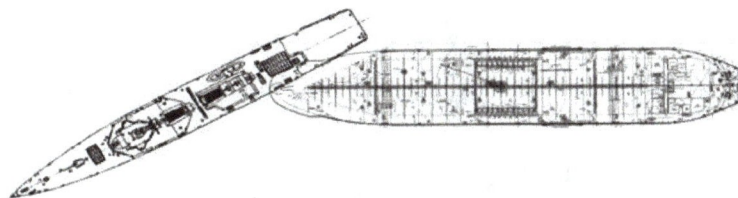

**Figure 7 – Approximate Geometry and Point of Impact
between USS JOHN S MCCAIN and ALNIC MC**

Despite their close proximity, neither JOHN S MCCAIN nor ALNIC sounded the five short blasts of whistle required by the International Rules of the Nautical Road for warning one another of danger, and neither attempted to make contact through Bridge to Bridge communications.

3. IMPACT OF THE COLLISION

The bulbous bow of ALNIC MC impacted JOHN S MCCAIN on the port (left) aft side. The impact created a 28-foot diameter hole both below and above the waterline of the JOHN S MCCAIN. See Figures 7, 8, and 9.

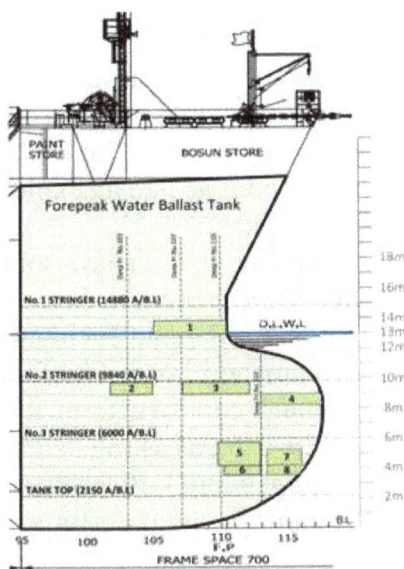

Figure 8 – Bulbous bow of ALINIC MC and damage to hull from bow to stern

Figure 9 – Point of impact on JOHN S MCCAIN from ALINIC MC

The point of impact was centered on Berthings 3 and 5 as noted in Figure 9. All significant injuries occurred to Sailors that were in Berthing 3 at the time of the impact. All ten of the fallen Sailors were in Berthing 5 at the time of impact.

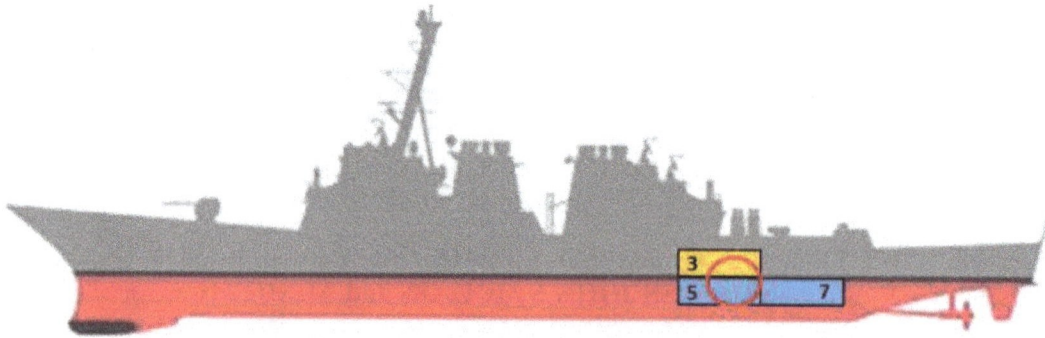

Figure 10 - Depiction of Approximate Location of Point of Impact

ALNIC MC and JOHN S MCCAIN initially remained attached to each other after the collision. Sailors describe this as lasting up to a couple of minutes. The prolonged contact kept the ship from taking a list (tilt to one side) immediately. Sailors on the bridge and on the external deck of the ship immediately after the collision could see ALNIC MC's bow (front of the ship) still lodged into the side of JOHN S MCCAIN. However, within 15 minutes JOHN S MCCAIN had developed a four degree list to port as the ship flooded.

The collision was felt throughout the ship. Watchstanders on the bridge were jolted from their stations momentarily and watchstanders in aft steering were thrown off their feet. Several suffered minor injuries. Some Sailors thought the ship had run aground, while others were concerned that they had been attacked. Sailors in parts of the ship away from the impact point compared it to an earthquake. Those nearest the impact point described it as like an explosion.

As required by Navy procedures, the Executive Officer ordered the collision alarm sounded to alert personnel to begin damage control efforts. The Commanding Officer remained on the bridge and the Executive Officer departed to the Combat Information Center and eventually to Berthing 3 to provide oversight in damage control efforts. The Command Master Chief, the senior assigned enlisted Sailor onboard, went to the area where damage control efforts, known as the Central Control Station, were managed and then moved about the ship, assisting damage control efforts. After the situation on the bridge stabilized, the Commanding Officer then proceeded to Central Control Station to check on the status of the damage control efforts.

The CO ordered the watch team to announce the collision on the Bridge-to-Bridge radio, which alerted other ships in the area to the collision and the damages. At 0530, JOHN S MCCAIN requested tugboats and pilots from Singapore Harbor to assist in getting the ship to Changi Naval Base.

JOHN S MCCAIN changed its lighting configuration at the mast to one red light over another red light, known as "red over red," the international lighting scheme that indicates a ship that is "not under command." Under the International Rules of the Nautical Road, this warns other ships that, due to an exceptional circumstance, a vessel is unable to maneuver as required.

Most of the electronic systems on the bridge were inoperable until the two ships parted. Main communications systems on the bridge stopped working after the collision and the bridge began using handheld radios to communicate with aft steering. Sound powered phones, which do not require electrical power to transmit communications, and handheld radios were the main means of communication from the bridge. Aft Internal Communications, a space adjacent to Berthing 5 with communications control equipment, quickly flooded and was likely responsible for the loss of bridge communications.

All U.S. Navy ships are designed to withstand and recover from damage due to fire, flooding, and other damage sustained during combat or other emergencies. Each ship has a Damage Control Assistant, working under the Engineering Officer, in order to establish and maintain an effective damage control organization. The Damage Control Assistant oversees the prevention and control of damage including control of stability, list, and trim due to flooding (maintaining the proper level of the ship from side to side and front and back), coordinates firefighting efforts, and is also responsible for the operation, care and maintenance of the ship's repair facilities. The Damage Control Assistant ensures the ship's repair party personnel are properly trained in damage control procedures including firefighting, flooding and emergency repairs. The Damage Control Assistant is assisted by the Damage Control Chief (DCC), a chief petty officer specializing in Damage Control. The officer in charge of damage control efforts, the Damage Control Assistant, called away General Quarters to notify the crew to commence damage control efforts.

General Quarters is a process whereby the crew reports to pre-assigned stations and duties in the event of large casualties such as flooding. General Quarters is announced by an alarm that sounds throughout the ship to alert the crew of an emergency situation or potential combat operations. All crewmembers are trained to report to their General Quarters watch station and to set a higher condition of material readiness against fire, flooding, or other damage. This involves securing additional doors, hatches, scuttles, valves and equipment to isolate damage and prepare for combat. Navy crews train on Damage Control continuously, with drills being run in port and underway frequently to prepare the teams for damage to equipment and spaces. During any emergency condition (fire, flooding, combat operations), the Damage Control Assistant coordinates and supervises all damage control efforts from one of the three Damage Control Repair Lockers.

Damage Control Repair Lockers are specialized spaces stationed throughout the ship filled with repair equipment and manned during emergencies with teams of about 20 personnel trained to respond to casualties. There are three repair lockers on the JOHN S. MCCAIN: Repair Locker 2, Repair Locker 5, and Repair Locker 3. Repair Locker 2 covers the forward part of the ship, Repair Locker 5 covers the engineering spaces and Repair Locker 3 covers the aft part of the ship. Each locker is maintained with similar equipment. Personnel assigned to repair lockers are trained and qualified to respond to and repair damage from a variety of sources with a specific focus on fire and flooding. Each repair locker can act independently but is also designed to support the others and can take over the responsibilities for any locker if damage prevents that locker's use. The repair lockers are normally unmanned unless the ship sets a condition of higher readiness like General Quarters when they would be manned within minutes.

Sailors began to locate, report and track flooding, fire, and structural damage to the ship immediately. Significant damage was reported throughout the ship in the moments after the collision, including flooding, internal fuel leakage, loss of ventilation and internal communications, and degradation of many of the ship's other systems.

JOHN S MCCAIN began the process of accounting for all crew members immediately after the collision. This process continued even as the crew made emergency repairs, battled flooding, and helped each other out of damaged spaces. The damage control efforts made confirming the location of personnel difficult. Varying reports of missing Sailors were made in the first minutes after the collision. However, by the submission of the third complete report, there was reasonable confidence that the crew had been accounted for was correct because all of the ten missing Sailors had been consistently reported missing and all lived in Berthing 5, a space that was inaccessible and flooded.

3.1 Impact to Berthing 5

Berthing 5 is located aft (near the back of the ship) on the port side. See Figure 10. It is approximately 25 feet by 15 feet and has 18 racks, stacked as bunk beds three-high. Each row of racks has a locker for Sailors' belongings. There is a lounge with seats, a small table, and a wall-mounted television. There is a head with one toilet, one shower stall, and one sink.

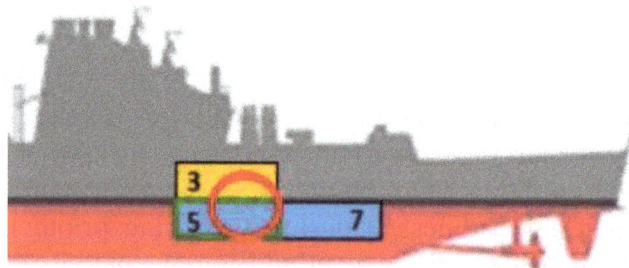

Figure 11 - Relative Positions of Berthings 3, 5 (in green), and 7, and point of impact

There are two means to exit Berthing 5: the primary egress (ladderwell) through a hatch with a scuttle (Figure 11) and an escape scuttle into Berthing 3 on the deck above (Figure 12).

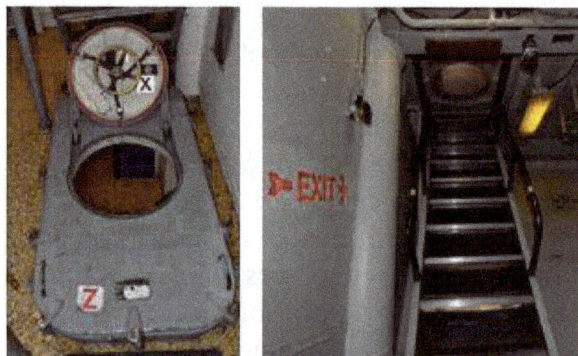

Figure 12 – Primary egress from Berthing 5
Left: from within Berthing 5
Right: Above on the deck outside Berthing 3

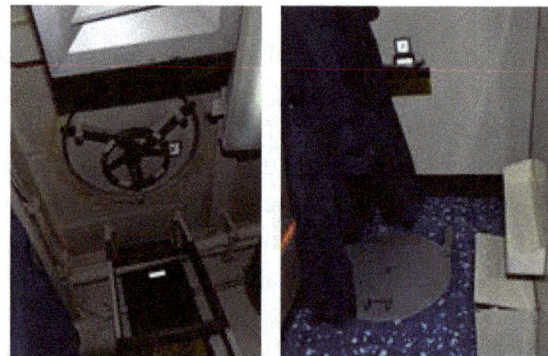

Figure 13 – Escape Scuttle from Berthing 5
Left: From within Berthing 5
Right: Above in Berthing 3

During Modified ZEBRA, the hatch is closed, but the scuttle is open for ease of access. The escape scuttle is normally closed at all times, as it was at the time of the impact. The collision knocked debris in Berthing 3 on top of the escape scuttle connecting Berthing 3 to Berthing 5 below it. This would have made any attempts to open and exit through the escape scuttle very difficult.

Most of the Berthing 5, a space that is normally 15 feet wide, was compressed by the impact to only 5 feet wide. There were 17 Sailors assigned to Berthing 5. At the time of the collision, all were aboard the ship and five were on watch or outside the space. Based on the size of the hole, and the fact that Berthing 5 is below the waterline, the space likely fully flooded in less than a minute after the collision.

Two Sailors who were in Berthing 5 at the time of the collision escaped from the space. The first Sailor was on the second step of the ladder-well leading to the deck above when the collision occurred. The impact of the collision knocked him to the ground, leaving his back and legs bruised. Fuel quickly pooled around him and he scrambled up and back onto the ladder. The Sailor climbed out of Berthing 5 through the open scuttle, covered in fuel and water from the near instantaneous flooding of the space. He did not see anyone ahead of or behind him as he escaped. He reported seeing two other Sailors in the lounge area, one preparing for watch duties and another standing near his rack. Both of these Sailors were lost, along with the eight shipmates who were in their racks to rest at the time of the collision.

The second Sailor who escaped from Berthing 5 heard the crashing and pushing of metal before the sound of water rushing in. Within seconds, water was at chest level. The passageway leading to the ladder-well was blocked by debris, wires and other wreckage hanging from the overhead. From the light of the battle lanterns (the emergency lighting that turns on when there is a loss of normal lights due to power outage) he could see that he would have to climb over the debris to get to the ladder-well.

As he started his climb across the debris to the open scuttle, the water was already within a foot of the overhead, so he took a breath, dove into the water, and swam towards the ladder-well. Underwater, he bumped into debris and had to feel his way along. He was able to stop twice for air as he swam, the water higher each time, and eventually used the pipes to guide him towards the light coming from the scuttle. The Sailor found that the blindfolded egress training, a standard that requires training to prepare Sailors for an emergency and was conducted when he reported to the command, was essential to his ability to escape.

One Sailor was alerted by the first Sailor who escaped Berthing 5 that others were still inside the space, and he went to assist them. When he first reached the closed hatch and open scuttle, the water in Berthing 5 was at the top of the third rung. He tried to enter the space, but was forced back up the ladder by the pressure of the escaping air and rising water, which within seconds had risen to within a foot of the hatch. He saw a Sailor swimming toward the exit and pulled him out of the water through the scuttle between the two decks. This was the second and last Sailor to escape from Berthing 5. His body was scraped, bruised, and covered with chemical burns from being submerged in the mixture of water and fuel.

An additional Sailor who came to assist observed the rescue and, looking down into the berthing, saw "a green swirl of rising seawater and foaming fuel" approaching the top of the scuttle. As the final Sailor to leave Berthing 5 was pulled to safety, the Sailors at the top of the scuttle checked to see if there was anyone behind him. They did not see anybody. By then, so much water was already coming up through the scuttle that it was difficult to close and secure. The fuel mixed in with the water made one of Sailor's hands so slippery that he cut himself while using the wrench designed to secure the scuttle, but the two were able to secure it to stop the rapid flooding of the ship.

3.2 Impact on Berthing 3

Berthing 3 is immediately above Berthing 5, but spans the width of the ship. There are two points of egress from each side of Berthing 3; on the port side there is a ladder-well leading down into the center of the berthing and an escape scuttle that is located in the forward section of the space leading up to the next deck. There were 71 Sailors assigned to Berthing 3.

At 0530, the DCA began receiving reports of a ruptured fire main and water and fuel flooding into Berthing 3. The port side of Berthing 3 suffered substantial damage, including a large hole in the bulkhead. See Figure 13. Racks and lockers detached from the walls and were thrown about, leaving jagged metal throughout the space. Cables and debris hung from the ceiling.

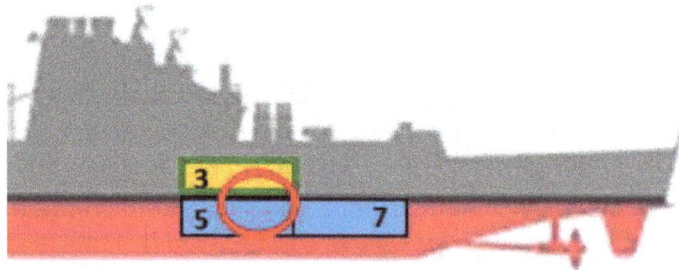

Figure 14 - Relative Positions of Berthings 3, 5, and 7 and point of impact

A Sailor from Berthing 3, who was later medically evacuated from the ship, sustained his injuries as the wall next to him blew apart in the collision and threw him to the ground. Water and fuel quickly pooled around him in the short time he was on the ground, and he began crawling over debris to escape. Another Sailor went to him and helped pull him to the lounge area and toward the ladder. On the way, the Sailor who was being assisted fell on the slippery floor and hit his head. Two other Sailors, also injured, helped him reach the flight deck.

Limited lighting guided the remaining Sailors as they left the berthing space. Sailors had to climb over lockers and other debris to escape, using the high vantage point to also minimize the risk of electrocution from traveling through the rising water. Some escaped in only their underwear, and many were bruised and bloodied from injuries sustained in the collision and covered in fuel. At least one Sailor attempted to move the metal rack pinning a trapped shipmate, and realized that he could not move it alone. The Sailors who escaped Berthing 3 provided some of the first reports to CCS that the space was severely damaged, that it was rapidly taking on water, and that Sailors were trapped inside.

Hearing reports that Sailors were trapped in Berthing 3, the Executive Offer and a group of Sailors, including some who evacuated Berthing 7, went to check on their shipmates. Several Sailors were pinned in their racks as a result of the collision, but, as the two ships pulled apart, the twisted metal shifted and most of the Sailors in Berthing 3 were able to escape as the debris moved. One of these Sailors was pinned in his rack underneath two racks that had collapsed and a number of lockers that became dislodged during the collision. He was able to escape after ALNIC MC detached. See Figures 14 and 15.

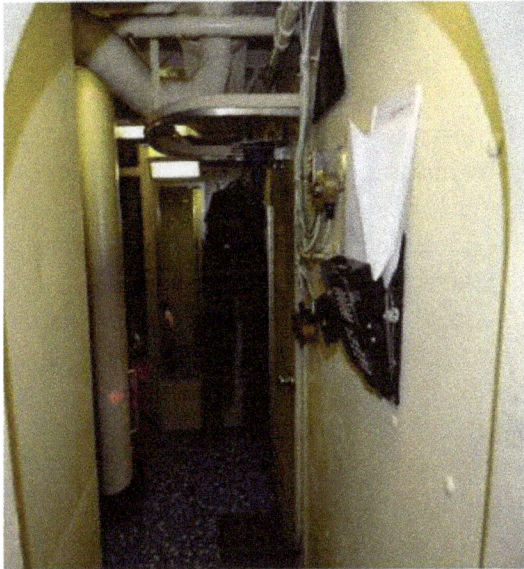

Figure 15 – Berthing 3 facing port

Figure 16 – Berthing 3 facing port after collision

However, two Sailors remained pinned in their racks even after the ships separated. Four members of the crew entered Berthing 3 through the jagged metal and rising water to rescue them. The first of these rescuers heard Sailors shouting for help from inside Berthing 3 and tried to enter on the port side; however, the door was blocked by debris, so he ran to the entrance on the starboard (right) side of the berthing.

One of the Sailors trapped in Berthing 3 had been asleep at the time of the collision and was awoken by it. When he opened his eyes, he understood that he was pinned in his rack, with one of his shoulders stuck between his rack and the rack above. He felt both air and water moving around him. He could hear shouting and began shouting himself, which alerted his others that he was trapped. Only his hand and foot were visible by those outside of the rack. The one battle lantern in the area provided the only light for rescuers to find the trapped Sailor. Water was already at knee level when rescuers reached him. The debris was too heavy for the rescuers to move, and a Portable Electric All-Purpose Rescue System, a "jaw of life" cutting device, was required to cut through the metal, separate the panels of the rack, and pull the panels out of the way. After approximately 30 minutes, these efforts allowed the trapped Sailor to pull his arm free. Moments later, the rescuers pulled him from between the racks by his foot. Stretcher bearers came to Berthing 3 and carried this Sailor to the Mess Decks to receive medical treatment.

The second Sailor was in a bottom rack in Berthing 3. His rack was lifted off the floor as a result of the collision, which likely prevented him from drowning in the rising water, and he was trapped at an angle between racks that had been pressed together. Light was visible through a hole in his rack and he could hear the water and smell the fuel beginning to fill Berthing 3. He attempted to push his way out of the rack, but every time he moved the space between the racks grew smaller and he was unable to escape. His foot was outside the rack and he could feel water. It was hot in the space and difficult to breathe, but he managed to shout for help and banged against the metal rack to get the attention of other Sailors in the berthing space. The Sailors who entered Berthing 3 to rescue others heard this and began assisting him, but he was pinned by more debris than the first Sailor freed.

It took approximately an hour from the time of the collision to free the second Sailor from his rack. Rescuers used an axe to cut through the debris, a crow bar to pull the lockers apart piece by piece, and rigged a pulley to move a heavy locker in order to reach the Sailor. Throughout the long process, his rescuers assured him by touching his foot, which was still visible. Once freed, the Sailor was the last person to escape Berthing 3. Everything aft of his rack was a mass of twisted metal. He had scrapes and bruises all over his body, suffered a broken arm, and had hit his head. He was unsure whether he remained conscious throughout the rescue.

At least one scuttle to Berthing 3 was shut during damage control efforts. The space was electrically isolated and, at 0608, the fire main valves were closed, reducing the amount of flooding. Dewatering efforts began and succeeded in removing the water from Berthing 3 prior to JOHN S MCCAIN's arrival at Changi Naval Base.

3.3 Impact on Berthings 4, 6, and 7

Berthings 5 and 7 are next to each other on the port side of the ship, mirrored by Berthings 4 and 6, respectively, on the starboard side of the ship. Berthings 4 and 5 are connected across the ship through "cross flooding ducts," designed to distribute water from port to starboard side (or vice versa) to keep the ship level if it takes on water. Berthings 6 and 7 are similarly connected. A six foot long crack in the wall between Berthings 5 and 7, created by the collision, allowed water to move between the spaces.

All Sailors in Berthing 7 were able to evacuate, but water was at approximately knee level as they exited the space. At 0530 there was report of a ruptured pipe in Berthing 7, which added to the flooding caused by the cracked wall separating Berthings 5 and 7. By 0605, Berthing 7 was reported as lost, meaning that it was fully flooded and secured to prevent the flooding from spreading to the rest of the ship.

All Sailors in Berthing 4 were able to evacuate. At 0544, Sailors reported 4 inches of water on the deck in Berthing 4. Sailors in Berthing 4 were thrown about their racks by the force of the collision. By 0627, the berthing space was lost. See Figures 16 and 17.

Figure 17—Scuttle and hatch into Berthing 4 showing the space completely flooded

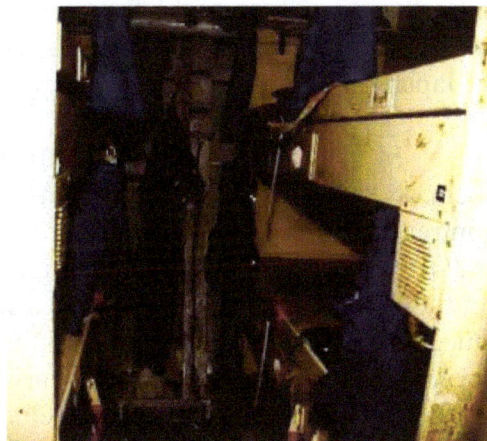

Figure 18—Berthing 4 racks after dewatering

All Sailors in Berthing 6 were able to evacuate. At 0546 flooding was reported in Berthing 6, which is across from Berthing 7 on the starboard side of the ship. Despite the crew's dewatering efforts, the space was declared lost at 0627.

At approximately 0630, as a result of crew's resiliency and successful damage control and engineering repair efforts, JOHN S MCCAIN was able to proceed under its own power toward Changi Naval Base, Singapore, at an average speed of 3 knots. JOHN S MCCAIN's navigation equipment was degraded as a result of the collision. While most electronic navigational aids on the bridge were operational, multiple warnings and alerts were illuminated, reducing the navigation team's confidence that the information was reliable. Because of the degraded information, the team relied on "seaman's eye" to stay on track while returning to port. Lack of ventilation across the ship raised concerns based on the amount of fuel that had spilled and the risks posed by rising temperatures inside the ship. The temperatures also drove many Sailors to the flight deck in order to escape the heat.

4. MEDICAL EFFORTS AND INJURIES

JOHN S MCCAIN medical teams established a triage center in Messing. This space, where the crew eats their meals, provided the largest open space on the ship and medical procedures can be performed on the cafeteria-style tables. The medical team treated lacerations and chemical burns from fuel exposure, splinted broken bones, and provided broad spectrum antibiotics to Sailors with open wounds. Triage care moved back to the Medical office at approximately 0630, as the initial rush of patients had been treated so the medical team would have full access to their equipment and supplies.

At approximately 0915, as the ship was transiting to Changi Naval Base, four seriously injured Sailors were medically evacuated to Singapore General Hospital by helicopter. Once pier-side at Changi Naval Base, another Sailor was transported to the hospital because of shock symptoms and an injury to his shoulder. This Sailor was one of the Sailors who had been trapped in Berthing 3. Three of the five medically evacuated Sailors were transported from Singapore to Yokosuka, Japan on 27 August 2017. The remaining two were transported back to Yokosuka, Japan on 28 August 2017.

As JOHN S MCCAIN approached Changi Naval Base, AMERICA approached alongside and two members of AMERICA's medical team came aboard to provide additional support, including intravenous fluids to treat heat stroke. Once the ships were pier side, AMERICA hosted the JOHN S MCCAIN medical team, together treating Sailors with cuts and chemical burns from fuel exposure. Until alternative arrangements could be made, AMERICA also provided meals for all JOHN S MCCAIN Sailors and berthing for over 150 Sailors whose berthings were flooded. The Sailors and Marines aboard AMERICA also provided initial support for the JOHN S MCCAIN crew, including daily supplies, watchstanders, counseling, and communications network support.

In total, 48 Sailors suffered injuries that required medical treatment. Five Sailors who were treated at Singapore General Hospital suffered severe injuries and were unable to return to their duties for more than 24 hours. Of the 48 injured Sailors, 43 continued to assist with damage control and recovery efforts immediately following the collision.

5. SEARCH AND RESCUE EFFORTS - 21 TO 24 AUGUST 2017

Though the ship did not have a complete muster confirming ten Sailors were missing until 0715, JOHN S MCCAIN reported that Sailors were believed missing within moments of the collision. Coordination began immediately for search and rescue (SAR) efforts in the water space surrounding the collision site.

At approximately 0715 on 21 August 2017, SAR Operations commenced with Commander, Amphibious Squadron 3 (CPR 3) as SAR On-Scene Commander. At approximately 0700, AMERICA was en route and preparing to launch MV-22B Ospreys and MH- 60S Sea Hawks to support SAR efforts once in range. Republic of Singapore Navy (RSN) and Republic of Singapore Coast Guard (RSNCG) SAR units were on station by 0800. Eventually there would be six Singaporean and six Malaysian vessels searching near the collision site.

At approximately 0900, the Republic of Singapore (RSN) deployed three ships with damage control equipment to assist and transfer equipment to JOHN S MCCAIN on a rigid-hulled inflatable boat (RHIB).

At approximately 1000 and 1030, two helicopters from AMERICA landed on the deck of JOHN S MCCAIN with damage control equipment and in support of the SAR efforts. By approximately 1400, U.S. Navy aircraft were conducting SAR efforts within 25nm of the collision point. A RSN helicopter, two RSN patrol boats, and a RSNCG vessel were on scene to assist.

The Malaysian Navy and RSN both searched 10nm on either side of the path that JOHN S MCCAIN had traveled, attempting to locate any Sailors that may have fallen through the hole in the ship's hull made by the collision. Throughout the evening of 21 August 2017, and continuously until 2000 on 24 August 2017, aircraft and surface vessels from the U.S. Navy, RSN, RSNCG, Singapore Air Force, Singapore Maritime Port Authority, Royal Malaysian Navy, Malaysian Maritime Enforcement Agency, Indonesian Navy and Royal Australian Air Force conducted multinational SAR operations. These efforts were coordinated from AMERICA, lasting for more than 80 hours and spanning more than 2,100-square miles.

On 22 August 2017, a body was found in the water by Malaysian units assisting the SAR efforts. The body was determined not to be one of the Sailors missing from JOHN S MCCAIN. SEVENTH Fleet suspended all SAR efforts outside the hull of JOHN S MCCAIN at sunset on 24 August 2017. Recovery efforts inside the hull of the ship continued.

6. DIVING AND RECOVERY OPERATIONS

A team of Navy Divers arrived on JOHN S MCCAIN as the ship entered the harbor in Singapore at approximately 1200 on 21 August 2017. They began inspecting the ship to determine how best to proceed with a dive inside the ship. The leader of the dive team inspected Berthing 3 and saw waves breaking into the ship. The divers discovered the hole in the port side of JOHN S MCCAIN that was approximately 28 feet wide. See Figure 18.

Figure 19 – Port side of JOHN S MCCAIN post-collision

By approximately 1435, JOHN S MCCAIN was moored and divers were in the water looking for places to enter the hull of the ship. The hole in the port side penetrated not only the hull, but an internal fuel tank as well. The fuel in the water created a number of hazards to divers and required them to proceed cautiously.

On a second dive at approximately 1500, divers were able to enter the hull of the ship to do initial safety assessments. Many of the conditions they found led to a cautious approach to assure the safety of the divers. The large amount of debris and structural damage required the divers to move slowly about the ship, even cutting holes through racks to access parts of the space. Visibility in Berthing 5 was very poor given the debris and lack of light. The divers had to move about the space almost exclusively by feel. The dive team conducted nearly continuous dive operations over a period of seven days until all ten of the Sailors in Berthing 5 were recovered.

7. FINDINGS

Collisions at sea are rare and the relative performance and fault of the vessels involved is an open admiralty law issue. The Navy is not concerned about the mistakes made by ALNIC. Instead, the Navy is focused on the performance of its ships and what we could have done differently to avoid these mishaps.

In the Navy, the responsibility of the Commanding Officer for his or her ship is absolute. Many of the decisions made that led to this incident were the result of poor judgment and decision making of the Commanding Officer. That said, no single person bears full responsibility for this incident. The crew was unprepared for the situation in which they found themselves through a lack of preparation, ineffective command and control and deficiencies in training and preparations for navigation.

7.1 Training

From the time when the CO ordered the Helm and Lee Helm split, to moments just before the collision, four different Sailors were involved in manipulating the controls at the SCC.

Because steering control was in backup manual at the helm station, the offer of control existed at all the other control stations (Lee Helm, Helm forward station, Bridge Command and Control station and Aft Steering Unit). System design is such that any of these stations could have taken control of steering via drop down menu selection and the Lee Helm's acceptance of the request. If this had occurred, steering control would have been transferred.

When taking control of steering, the Aft Steering Helmsman failed to first verify the rudder position on the After Steering Control Console prior to taking control. This error led to an exacerbated turn to port just prior to the collision, as the indicated rudder position was 33 degrees left, vice amidships. As a result, the rudder had a left 33 degrees order at the console at this time, exacerbating the turn to port.

Several Sailors on watch during the collision with control over steering were temporarily assigned from USS ANTIETAM (CG 54) with significant differences between the steering control systems of both ships and inadequate training to compensate for these differences.

Multiple bridge watchstanders lacked a basic level of knowledge on the steering control system, in particular the transfer of steering and thrust control between stations. Contributing, personnel assigned to ensure these watchstanders were trained had an insufficient level of knowledge to effectively maintain appropriate rigor in the qualification program. The senior most officer responsible for these training standards lacked a general understanding of the procedure for transferring steering control between consoles.

7.2 Seamanship and Navigation

Much of the track leading up to the Singapore Traffic Separation Scheme was significantly congested and dictated a higher state of readiness. Had this occurred, maximum plant reliability could have been set with a Master Helmsman and a qualified Engineering Lee Helm on watch.

If the CO had set Sea and Anchor Detail adequately in advance of entering the Singapore Strait Traffic Separation Scheme, then it is unlikely that a collision would have occurred. The plan for setting the Sea and Anchor Detail was a failure in risk management, as it required watch turnover of all key watch stations within a significantly congested TSS and only 30 minutes prior to the Pilot pickup.

If JOHN S MCCAIN had sounded at five short blasts or made Bridge-to-Bridge VHF hails or notifications in a timely manner, then it is possible that a collision might not have occurred.

If ALNIC had sounded at least five short blasts or made Bridge-to-Bridge VHF hails or notifications, then it is possible that a collision might not have occurred.

7.3 Leadership and Culture

The Commanding Officer decided not to station the Sea and Anchor detail when appropriate, despite recommendations from the Navigator, Operations Officer and Executive Officer.

Principal watchstanders including the Officer of the Deck, in charge of the safety of the ship, and the Conning Officer on watch at the time of the collision did not attend the Navigation Brief the afternoon prior. This brief is designed to provide maximum awareness of the risks involved in the evolution.

Leadership failed to provide the appropriate amount of supervision in constructing watch assignments for the evolution by failing to assign sufficient experienced officers to duties.

The Commanding Officer ordered an unplanned shift of thrust control from the Helm Station to the Lee Helm station, an abnormal operating condition, without clear notification.

No bridge watchstander in any supervisory position ordered steering control shifted from the Helm to the Lee Helm station as would have been appropriate to accomplish the Commanding Officer's order. As a result, no supervisors were aware that the transfer had occurred.

Senior officers failed to provide input and back up to the Commanding Officer when he ordered ship control transferred between two different stations in proximity to heavy maritime traffic.

Senior officers and bridge watchstanders did not question the Helm's report of a loss of steering nor pursue the issue for resolution.

This assessment of USS John S. McCain is not intended to imply that ALNIC mistakes and deficiencies were not also factors in the collision.

ANNEX A - TIMELINE OF EVENTS

20 August 2017

1300	Navigation Brief to prepare the crew for the Singapore Strait transit and entering Sembawang, Singapore.
Approx. 1326	Rudder swing checks were completed verifying satisfactory operation of the rudder.
1730	The Commanding Officer retired to his cabin to rest before reporting to the bridge at 0115 the next morning.
1904	JOHN S MCCAIN energized Navigation Lights.
2115	Modified Condition Zebra was verified. As explained in the report, this condition maximizes the ability of the ship to gain a watertight status in the event of collision.

21 August 2017

0000	JOHN S MCCAIN is en route Singapore.
0001	Log entries reported that one surface search radar was non-operable.
Approx. 0100-0101	Navigation watchstanders began to verify the ship's position at more frequent intervals (15 minutes).
0115	The Commanding Officer arrived on the Bridge.
Between approximately 0127 & 0204	Key supervisory watch stations changed personnel.
0216	Watchstanders shifted propulsion operations to what is termed split plant, a condition in which different gas turbines drive each of the two shafts separately.
0300	Currents were running at a speed of 2.7 knots requiring steering adjustment.
0315	Watchstanders report visual sighting of land.
0418	Additional watchstanders reported for duties as the Modified Navigation Detail.
0426	Navigation watchstanders began determining the ship's position at more frequent intervals (5 minutes).
0427	JOHN S MCCAIN turned to avoid surface contacts in the area.

0430	The Executive Offer arrived on the bridge.
0436	The Commanding Officer ordered steering modes shifted from automatic control to backup manual control.
Approx. 0436	Personnel responsible for tracking contacts on radar secured the auto-tracking feature on the SPS-67 radar and began manually tracking surface contacts.
Starting at 0437	The bridge ordered various rudder orders to avoid shipping. None of these maneuvers were logged.
0444	JOHN S MCCAIN turned to port and steadied on course 227T. On this course, the ship was aligned to enter the westbound Singapore Strait Traffic Separation Scheme.
Approx. 0454	Radar contact was gained on the ALNIC nearly ahead of JOHN S MCCAIN on the port side, within 8 nautical miles. ALNIC was in the center of a group of three other contacts traveling in the same general direction as JOHN S MCCAIN. Watchstanders did not discuss maneuvering intentions with respect to these contacts.
Approx. 0457	JOHN S MCCAIN increased speed to 17 knots.
0459	JOHN S MCCAIN reduced speed to 16 knots.
0500	Reveille was announced to wake the crew for entering port. The Navigator informed the OOD that previous course changes to the North to avoid surface traffic had put JOHN S MCCAIN behind on its intended track and timeline and recommended an increase in speed to make 18 knots.
0500 – 0524	JOHN S MCCAIN overtook several vessels just north of the eastern entrance to the Singapore Strait Traffic Separation Scheme. The closest point of approach during these passages was as close as 600 yards.
0509	JOHN S MCCAIN altered course to 226T.
0513	JOHN S MCCAIN increased speed to 18 knots and was steady on course 226T.
0514	JOHN S MCCAIN increased speed to 20 knots and was steady on course 226T.
0518	JOHN S MCCAIN turned starboard to course 230T, speed 20 knots. The Helmsman was compensating for the effects of currents with between 1 - 4 degrees of right rudder to stay on course 230T.
Approx. 0519	The Commanding Officer noticed the Helmsman was struggling to maintain course while simultaneously adjusting throttles. The CO ordered steering

control separated from propulsion control, with duties divided between the Helm and Lee Helm watch stations. Splitting of the Helm and Lee Helm was not previously discussed at the Navigation Brief or at any time prior to the CO ordering it.

Approx. 0520	Supervisory watch stations reported that the Automated Identification System (AIS) representation of contacts was cluttered and "useless." Commercial traffic routinely reports positions via this system, enabling other vessels to use Global Positioning System (GPS) satellite information to accurately determine their positions.
05:20:03	JOHN S MCCAIN was overtaking motor vessel GUANG ZHOU WAN. JOHN S MCCAIN was making 18.6 knots over ground. JOHN S MCCAIN closed range from behind ALNIC on ALNIC's starboard side.
0520:39	The Lee Helm station took control of steering in computer assisted mode. The shift in steering locations caused the rudder to move amidships.
0520:47	Lee Helm took control of the port shaft. Port and starboard shafts were both at 087 RPM/100.1% pitch.
Just before 0521	The Helm reported to his immediate supervisor that he had lost steering control. The supervisor informed the Helm to inform the officer in charge of ship safety and navigation, the Officer of the Deck.
0521	The Helm reported loss of steering to the Officer of the Deck. The rudders were amidships. JSM was on course 228.7T, engines were all ahead full for 20 knots. JSM was making 18.6 knots over ground and turning to port at 0.26 degrees per second. ALNIC was on course 230T, speed 9.6 knots, and was bearing 164T at a range of approximately 582 yards from JSM.
0521	The Conning Officer, the person responsible for issuing steering orders, ordered the Helm to shift steering control to the offline steering units, 1A and 2A.
0521	A loss of steering casualty on the ship's general announcing circuit was announced and After Steering was ordered manned. After Steering is an auxiliary station that has the ability to take control of steering in the event of a problem or casualty to the ship's primary control stations.
0521:13	Steering units on the port rudder were shifted as ordered.
0521:15	Steering units on the starboard rudder were shifted as ordered.
0521:55	The first watchstander reported to After Steering. JOHN S MCCAIN did not have a complete delineated list of personnel to man After Steering in the event of a casualty or problem.

64

0522	JOHN S MCCAIN was on course 216.3T, speed 18.4 knots and was turning to port at a rate of approximately 0.2 degrees per second. Bridge watchstanders followed the Commanding Officer's order to change the lighting configuration to indicate a vessel not under command by the International Rules of the Nautical Road.
Approx. 0522:04	The Lee Helm took control of the starboard shaft. The port and starboard shafts remained at a speed of 087 RPM and 100.1% pitch. The Lee Helm did not match the port and starboard throttles that control the speed of the shafts. JOHN S MCCAIN was on course 216.1T and turning to port at a rate of approximately 0.25 degrees per second. Rudders were amidships.
Approx. 0522:05	The Commanding Officer ordered the ship slowed with a reduction in speed to 10 knots.
0522:07	The command to the port shaft lowered speed to 44 RPM and 100.1% pitch. The starboard shaft remained at a speed of 87 RPM and 100.1%. Rudders were amidships. No bridge watchstanders were aware of the mismatch in thrust and the effect on causing the ship's turn to port.
0522:40	JOHN S MCCAIN was on course 204.4T, speed 16.6 knots and was turning to the left at a rate of approximately 0.41 degrees per second.
0522:45	The Executive Officer noticed the ship was not slowing down as quickly as expected and alerted the Commanding Officer. In response, the Commanding Officer ordered 5 knots. This order was echoed by the Conning Officer. The CO did not announce that he had taken direct control of maneuvering orders as required by Navy procedures.
0523:00	The Conning Officer ordered right standard rudder. JOHN S MCCAIN was on course 194.5T at a speed of 15.8knots. ALNIC was on course 229.8T, 9.6 knots, and was bearing 097T from JOHN S MCCAIN at a range of approximately 368 yards.
0523:01	After Steering took control of steering in backup manual mode.
0523:06	The port shaft continued to slow. The starboard shaft was ahead at a speed of 87 RPM and 100.1% pitch. The port shaft order at this time was 32 RPM at 81.1% pitch. JSM was on course 192T, speed 15.6 knots and turning to the left at a rate of approximately 0.5 degrees per second.
0523:16	The Helm took control of steering at the helm station in Backup Manual mode.
0523:24	Throttles were finally matched at the Lee Helm station and both shafts were ahead to reach 5 knots. JOHN S MCCAIN was on course 182.8T, speed 13.8 knots, and turning to port at a rate of approximately 0.54 degrees per second.

0523:27	Aft Steering Helmsman took control of steering. This was the fifth transfer of steering and the second time the Aft Steering unit had gained control in the previous two minutes.
0523:44	JOHN S MCCAIN was on course 177T, speed 11.8 knots, and was slowly turning to the left port at a rate of approximately .04 degrees per second. The ordered and applied right 15 degree rudder checked JOHN S MCCAIN's swing to port and the ship was nearly on a steady course.
0523:58	ALNIC's bulbous bow struck JSM between frame 308 and 345 and below the waterline.
0524:12	After Steering still had control of steering at the ASU in CAM but the rudders moved amidships.
0524:24	JSM engines answered "all stop" and the shafts came to idle speed. The ship was on course 138.6T, speed 5.7 knots, and the ship was turning to port at a rate of approximately 1.4 degrees per second.
0526	JSM set General Quarters and the Damage Control Assistant assumed responsibility for all DC efforts from CCS.

B1: Port side view of JOHN S MCCAIN post-collision

B2: Internal and external views of the damage caused by the collision

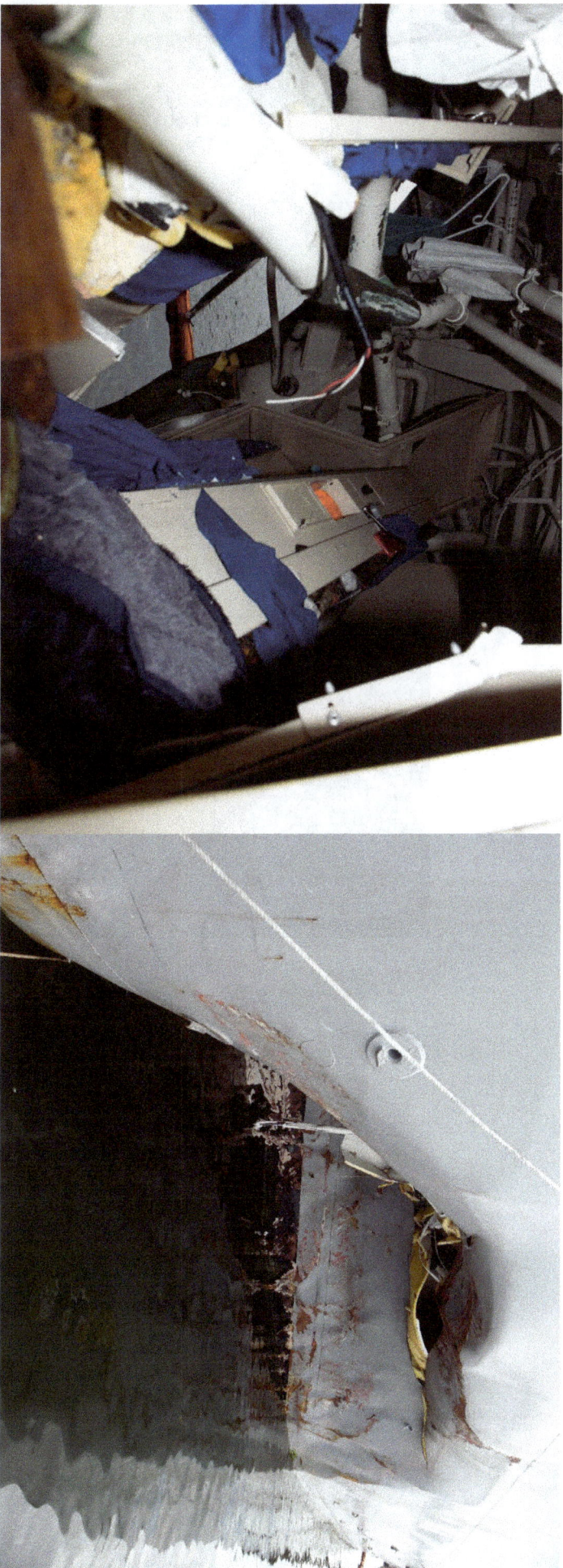

Internal view through Berthing 3 to point of impact; ocean visible

External view of point of impact; rupture in hull of the ship visible

B3: Berthing 3 Primary egress (ladderwell) of JOHN S MCCAIN post-collision, with views into Berthing 3

View through hatch into Berthing 3

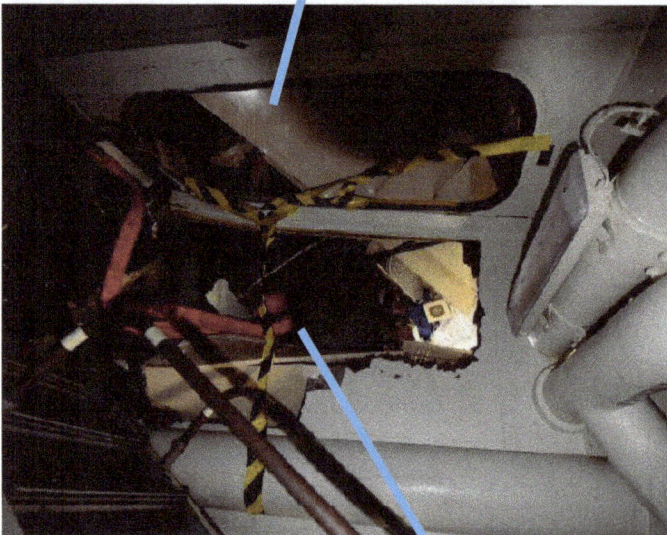

View through hole cut into bulkhead (wall)

B4: View within Berthing 3, facing port (comparison)

Undamaged Flight 1 Arleigh Burke Class Destroyer

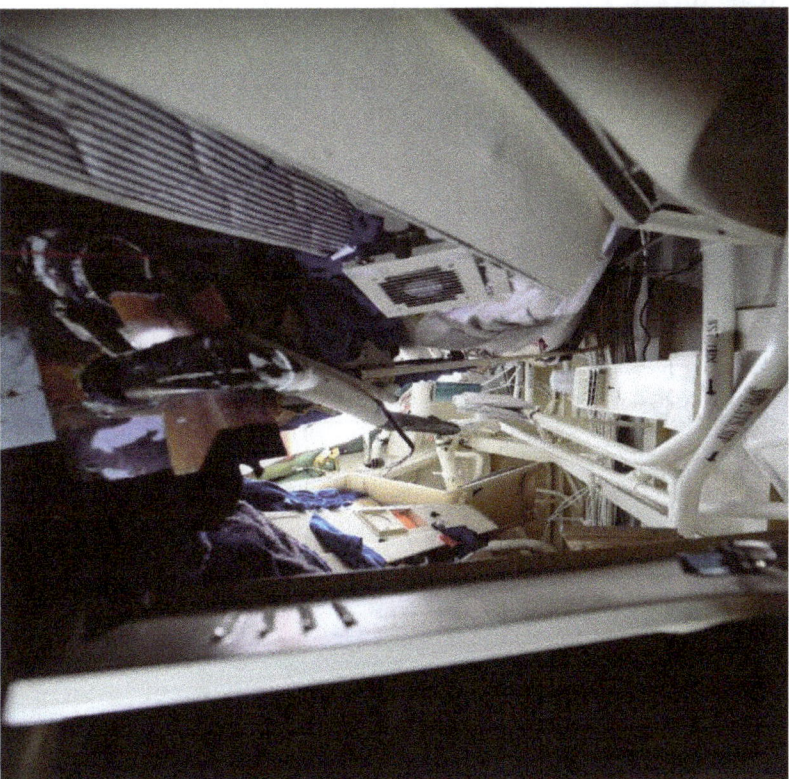

JOHN S MCCAIN, post-collision

ANNEX C – Catalog of Flooding in Spaces aboard USS JOHN S MCCAIN

COMPARTMENT	FRAME NO.	FLOODING
Crew Living No. 3	2-300-01-l	2 Feet
Physical Fitness RM	2-300-2-L	2 feet
Access TK	2-305-2-T	2 Feet
Crew WR, WC, & SH	2-321-2-L	2 Feet
Access TK	2-326-1-T	2 Feet
Crew WR, WC, & SH	2-326-0-L	2 Feet
Crew Library	2-338-2-L	Little
Power CONV Room	3-319-0-Q	Partial
MER	4-254-0-E	Minor
IC & Gyro	3-300-0-C	SOLID
Crew Living No. 4	3-300-1-L	SOLID
Crew WR, WC, & SH	3-300-2-L	SOLID
Crew Living No. 5	3-310-2-L	SOLID
Crew WR, WC, & SH	3-325-1-L	SOLID
Crew Living No. 6	3-338-1-L	SOLID
Crew Living No. 7	3-338-2-L	SOLID
CG Locker	3-338-3-A	SOLID
Crew WR, WC, & SH	3-338-5-L	SOLID
CG Locker	3-338-4-A	SOLID
Crew WR, WC, & SH	3-338-6-L	SOLID
VCHT RM No. 2	3-300-0-E	SOLID
A/C Mach & PMP Rm	5-300-01-E	SOLID
Fuel Service Tank	5-300-4-F	Compromised
Cross Flooding Ducts	FR 335 & 367	SOLID
Fuel RCVG TK	5-338-2-F	Suspected Flooded